Sparse Matrices

This is Volume 99 in
MATHEMATICS IN SCIENCE AND ENGINEERING
A series of monographs and textbooks
Edited by RICHARD BELLMAN, *University of Southern California*

The complete listing of books in this series is available from the Publisher
upon request.

SPARSE MATRICES

REGINALD P. TEWARSON

Department of Applied Mathematics and Statistics
State University of New York
Stony Brook, New York

ACADEMIC PRESS   New York and London   1973

ACADEMIC PRESS, INC.
111 Fifth Avenue, New York, New York 10003

United Kingdom Edition published by
ACADEMIC PRESS, INC. (LONDON) LTD.
24/28 Oval Road, London NW1

LIBRARY OF CONGRESS CATALOG CARD NUMBER: 72-88359

AMS (MOS) 1970 Subject Classifications: 65F05, 65F15, 65F25, 65F30

PRINTED IN THE UNITED STATES OF AMERICA

To

HEIDI, ANITA, AND MONIQUE

Contents

Chapter 3. **Additional Methods for Minimizing the Storage for EFI**

Chapter 4. **Direct Triangular Decomposition**

Chapter 5. **The Gauss–Jordan Elimination**

Chapter 6. **Orthogonalization Methods**

Contents

Chapter **7.   Eigenvalues and Eigenvectors**

Chapter **8.   Change of Basis and Miscellaneous Topics**

Preface

The purpose of this book is to present in a unified and coherent form the large amount of research material currently available in various professional journals in the area of computations involving sparse matrices. At present, results in this area are not available in book form, particularly those involving direct methods for computing the inverses and the eigenvalues and eigenvectors of large sparse matrices.

Sparse matrices occur in the solution of many important practical problems, e.g., in structural analyses, network theory and power distribution systems, numerical solution of differential equations, graph theory, as well as in genetic theory, behavioral and social sciences, and computer programming. As our technology increases in complexity, we can expect that large sparse matrices will continue to occur in many future applications involving large systems, e.g., scheduling problems for metropolitan fire departments and ambulances, simulation of traffic lights, pattern recognition, and urban planning.

My interest in the area of sparse matrices dates back to 1962–1964, when I helped design and write a computer code for solving linear programming problems for a major computer manufacturer. The matrices occurring in linear programming problems are generally large and sparse (they have few nonzero elements); therefore, in order to make the code efficient, only the nonzero elements of such matrices are stored and operated on. I found then that very little published material was available on computing the sparse factored form of inverses needed in the linear programming algorithms. This experience led to the publication of a number of research papers.

In the spring of 1968, I was invited to speak at a Symposium on Sparse Matrices and Their Applications held at IBM, Yorktown Heights, New York in September of the same year. Another invitation followed in 1969 to present a paper at a Conference on Large Sparse Sets of Linear Equations at Oxford University in April 1970. In the summer of 1969, I wrote a survey paper on computations with sparse matrices at the request of the editors of the Society of Industrial and Applied Mathematics, which appeared in the September 1970 issue of the SIAM Review. In the same year Professor R. Bellman suggested that I write a book on this subject. It was a happy coincidence when Professor L. Fox asked me to give a graduate seminar on Sparse Matrices during the Hillary term 1970 at Oxford University. Out of these lectures the book grew.

This book is intended for numerical analysts, computer scientists, engineers, applied mathematicians, operations researchers, and others who have occasion to deal with large sparse matrices. It is aimed at graduate–senior level students. It is assumed that the reader has had a course in linear algebra. I have tried to avoid a terse mathematical style at the cost of being at times redundant and attempted to strike a balance between rigor and application. I believe that applications should lead to generalizations and abstractions. As far as is possible, the algorithmic or constructive approach has been followed in the book.

I have given the basic techniques and recent developments in direct methods of computing the inverses and the eigenvalues and eigenvectors of large sparse matrices. I have avoided including material which is readily available in well-known texts in numerical analysis, except that needed as a basis for the material developed in this book.

The organization of the text is as follows.

In Chapter 1, several commonly used schemes for storing large sparse matrices are described, and a method for scaling matrices is given, such that in computations involving the scaled matrices, the round-off errors remain small.

In Chapter 2, a discussion of the well-known Gaussian elimination method is given. It is shown how the Gaussian elimination can be used to express the inverse of a given sparse matrix in a factored form called the Elimination Form of Inverse (EFI). Techniques are given for getting as sparse an EFI of a given sparse matrix as possible. Some methods for minimizing the total number of arithmetical operations in the evaluation of the EFI are also described. The storage and the use of the EFI in practical computations are discussed.

In Chapter 3, several methods are given for obtaining a reasonably sparse EFI. These methods do not require as much work as those in Chapter 2. The permutation of the given sparse matrix to one of the several forms (e.g., the band form) that are desirable for getting a sparse EFI is also discussed.

The Crout, Doolittle, and Choleskey methods, which are closely related to the Gaussian elimination method, are considered in Chapter 4. Techniques for minimizing the number of nonzeros created at each step for these methods are given; these techniques are naturally similar to those given in Chapters 2 and 3 for the Gaussian elimination method.

In Chapter 5, the well-known Gauss–Jordan elimination method is investigated, and it is shown how another factored form of inverse, called the Product Form of Inverse (PFI) can be obtained. The relation between the PFI and the EFI, as well as the techniques for finding a sparse PFI are also given.

The orthonormalization of a given set of sparse vectors by using the Gram–Schmidt, the Householder, or the Givens method is discussed in Chapter 6. The last two methods are also used in Chapter 7 for evaluating the eigenvalues and eigenvectors of sparse matrices. Another method in Chapter 7 makes use of a technique similar to the Gaussian elimination method to transform the given matrix. In both chapters, techniques are described which tend to keep the total number of new nonzeros (created during the computational process) to a minimum.

Finally, in Chapter 8, the relevant changes in the EFI or the PFI, when one or more columns of the given matrix are changed, are described. This happens in many applications, e.g., linear programming. Another factored form of the inverse, which is similar to the EFI, is also given.

A comprehensive bibliography on sparse matrices follows Chapter 8.

Acknowledgments

I would like to express my thanks to the following: Professor L. Fox, for inviting me to spend my sabbatical year at Oxford University, where most of this book was written; Professor R. Bellman, for his initiative in suggesting that this book be written, as well as for his advice and encouragement; Professor R. Joseph and my Ph.D. students—Mr. I. Duff (Oxford), Mr. Y. T. Chen, and Mr. K. Y. Cheng —for reading and suggesting various improvements in the manuscript.

CHAPTER

I

Preliminary Considerations

1.1. Introduction

In this introductory chapter, we shall first mention some of the areas of application in which sparse matrices occur and then describe some commonly used schemes for storing such large sparse matrices in the computer (internal and/or external storage). A simple method of scaling matrices in order to keep round-off errors small is also given. The chapter ends with a bibliography and related comments.

1.2. Sparse Matrices

A matrix having only a small percentage of nonzero elements is said to be *sparse*. In a practical sense an $n \times n$ matrix is classified as sparse if it has order of n nonzero elements, say two to ten nonzero elements in each row, for large n. The matrices associated with a large class of man-made systems are sparse. For example, the matrix representing the communication paths of the employees in a large organization is sparse, provided that the ith row and the jth column element of the matrix is nonzero if and only if employees i and j interact. Sparse matrices appear in linear programming, structural analyses, network theory and power distribution systems, numerical solution of differential equations, graph theory, genetic theory, social and behavioral sciences, and computer programming.

The current interest in, and attempts at the formulation and solution of problems in the social, behavioral, and environmental sciences (in particular as such problems arise in large urban areas ; see, for example, Rogers, 1971) will in many cases lead to large sparse systems. If such systems are nonlinear, then their linearization—often the first step towards the solution—will result in still larger sparse systems.

Often, interesting and important problems cannot be solved because they lead to large matrices which either are impossible to invert on available computer storage or are very expensive to invert. Since such matrices are generally sparse it is useful to know the techniques currently available for dealing with sparse matrices. This allows one to choose the best technique for the type of sparse matrix he encounters. The time and effort required to develop the various techniques for handling sparse matrices is especially justified when several matrices having the same zero–nonzero structures but differing numerical values have to be handled. This occurs in many of the application areas already mentioned.

1.3.   Packed Form of Storage

Large sparse matrices are generally stored in the computers in *packed form*; in other words, only the nonzero elements of such matrices with the necessary indexing information are stored. There are four reasons for utilizing the packed form of storage. First, larger matrices can be stored and handled in the internal storage of the computer than is otherwise possible. Second, there are cases when the matrix even in packed form does not fit in the internal storage (for example, in time sharing) and external storage (for example, tapes or discs) must be used. Generally, getting the data from the external storage is much slower than internal computations involving such data, therefore, the packed form is preferred for the external storage also. Third, a substantial amount of time is saved if operations involving zeros are not performed; this is done by symbolic processing in which only the nontrivial operations are carried out. This is often the only way in which large matrices can be reasonably handled. Fourth, it turns out that the inverse of a given matrix expressed as a product of elementary matrices (only the nontrivial elements of such matrices are stored in packed form) usually needs less storage than the explicit inverse of the matrix in packed form. Such factored forms of inverses are particularly advantageous when they are later used for multiplying several row and column vectors, in linear programming, for example.

There are various packing schemes available, some of which are described below; these have been found efficient and are incorporated in computer codes.

Let A be a square matrix of order n with τ nonzero elements, where $\tau \ll n^2$, then A is clearly sparse. Let the ith row and the jth column element of A be denoted by a_{ij}. In order to store only the nonzero elements $a_{ij} \neq 0$, we need to store i, j and a_{ij}. If one cell of the storage is used for each of these quantities, then a total of 3τ cells will be needed to store all the nonzero elements of A. Evidently, 3τ should be substantially less than n^2 to make it worthwhile to spend the extra effort and computing time involved in packing.

In many algorithms that transform A to some other desirable form, additional nonzero elements are created in the various steps of the computations. Therefore, in the packed storage some provision has to be made to add new nonzero elements to the various columns (or rows) of A as the computation proceeds and the elements get changed. The ideal storage would be one which minimizes both the total storage used and the total computation time. In general, the two requirements, minimum storage and minimum time, are incompatible and a trade-off must be made.

UTILIZATION OF LINKED LISTS IN PACKING

One way of storing the nonzero elements of the given sparse matrix A is by making use of *linked lists* as follows. Each nonzero element a_{ij} is stored as an *item* in its column j (see Fig. 1.3.1). An item is an ordered

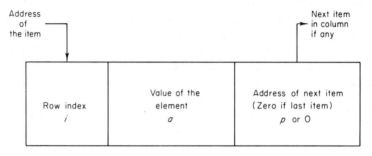

Fig. 1.3.1.   The item corresponding to a_{ij}.

triple (i, a, p), where i is the row index, a the value of the element a_{ij} and p is the address of the next nonzero element of column j. The address p is zero if the item corresponds to the last nonzero element of the column. The total storage consists of two parts, BC (*Beginning of Column address*) and SI (*Storage for Items*). The first part BC has n contiguous locations, each of which contains the starting address of the first item of the corresponding column. For example, the jth cell of BC has the starting address $SI(\alpha)$ of the item associated with the first nonzero element in column j (see Fig. 1.3.2). SI, the second part, consists of all items associated with the nonzero elements of A. Since A has τ nonzero elements and to each element there corresponds an item which is three

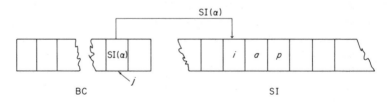

Fig. 1.3.2.   Linked lists packed storage.

elements long, SI will require 3τ storage locations, which need not be necessarily contiguous. Therefore, if we use linked lists, a total of $n + 3\tau$ storage locations is required to store the given matrix A in packed form.

The principal advantage of this storage scheme is that during the computations nonzero elements created in the columns can be easily stored in SI; there is no need to move down all the following elements, as would be the case in the usual storage scheme when a new element is inserted. Furthermore, the cells in SI need not be contiguous, as long as they are in groups which are divisible by 3.

Let us give a simple example showing how the creation of a new nonzero element affects BC and SI. Suppose, $a_{13} = a_{33} = 0, a_{23} = 0.5$, and $a_{43} = 1.5$; the storage for BC begins at location 101, and the items corresponding to a_{23} and a_{43} begin at locations 200 and 203 respectively. If later on a_{33} changes from zero to nonzero (say 2.5), and the corresponding item is to be stored starting at location 300, then the relevant changes can be exhibited as follows:

Location	103	200	201	202	203	204	300	301	302
Present contents	200	2	0.5	203	4	1.5	—	—	—
New contents	200	2	0.5	300	4	1.5	3	2.5	203

Thus, in the existing matrix storage only the contents of location 202 had to be modified in order to add a new nonzero element. However, if instead of a_{33}, a_{13} became nonzero (say 3.5) and the corresponding item was stored (as before) starting at location 300, then we would have the following.

Location	103	200	201	202	203	204	300	301	302
Present contents	200	2	0.5	203	4	1.5	—	—	—
New contents	300	2	0.5	203	4	1.5	1	3.5	200

In either case, it is evident that the contents of only one location in the original linked list have to be modified to insert a new nonzero element.

If during the computations some nonzero element becomes zero, then the storage so released by the corresponding item can be used for storing the items associated with new nonzero elements. The starting

addresses of such items which are available for storage can be main-
tained as a chained list by using the third cell of each item. Only the
address of the first available item storage has to be noted elsewhere.
The third cell of each available item storage contains the starting address
of the next available storage item. If it is the last available item storage,
then its third cell is zero. When a new item becomes available for
storage it is added to the top of the list. Similarly, available items from
the top of the list are used for storing new items.

Let us consider two simple examples to illustrate the above techniques. Suppose two items for storage were available, their starting
addresses were, respectively, 101 and 201, and we want to add another
available item storage starting at 301 to this list. If location 50 contains
the address of the first available item storage, then the appropriate
changes are shown below.

Location	50	101	102	103	201	202	203	301	302	303
Present contents	101	—	—	201	—	—	0	—	—	—
New contents	301	—	—	201	—	—	0	—	—	101

On the other hand, to store a new item, we use the first available item
storage in the above list, namely, locations 301, 302, and 303, and then
change the contents of the various locations in the preceding table to the
line labeled as present contents.

Sometimes methods of packing which do not use linked lists are
useful. They use less storage, but additional nonzero elements can be
introduced only by relocating all the succeeding elements, that is, by
moving them down. These schemes are suitable when only a small
portion of the matrix can be stored in the internal storage of the
computer at one time and a large amount of time would therefore be
required to transfer the data to and from the external storage. We
shall now describe four such schemes and show how a matrix A_5,
whose nonzero elements are $a_{21}, a_{41}, a_{52}, a_{13}, a_{33}, a_{24}$, and a_{45}, is
stored according to the first three schemes. The last scheme is for
symmetric matrices and therefore another matrix is utilized there. In
the first three schemes the matrix is stored by columns but in the last
one it is stored by rows.

SCHEME I

To each nonzero element of the matrix there corresponds an item of two storage cells. The first storage cell contains the row index and the second the value of the element. A zero row index in an item denotes the end of the current column. The second cell of such an item contains the index of the next column. Zeros in both cells of an item denote the end of the matrix storage. Thus there are $n + \tau + 1$ items in all, n for the columns, τ for the nonzero elements of A, and 1 to denote the end of the matrix storage. As each item uses two storage locations, a total of $2(n + \tau + 1)$ locations will be required to store A.

The matrix A_5 for which $\tau = 7$ and $n = 5$ is stored as the array

$(0, 1; 2, a_{21}; 4, a_{41}; 0, 2; 5, a_{52}; 0, 3; 1, a_{13}; 3, a_{33}; 0, 4; 2, a_{24}; 0, 5;$
$\quad 4, a_{45}; 0, 0).$

SCHEME II

The information about the given matrix is stored in three arrays: VE (Value of Elements), RI (Row Indices), and CIP (Column Index Pointer). $RI(\alpha)$, the αth element of RI, contains the row index of the corresponding element $VE(\alpha)$ of VE. If the first nonzero element of the βth column of the given matrix is in $VE(t_\beta)$ then t_β is stored in the βth element of CIP, namely, $CIP(\beta) = t_\beta$. It is evident that VE and RI each has τ elements but CIP has n elements. Thus $2\tau + n$ storage cells will be required in this scheme.

The storage for A_5 is as follows:

$$VE = (a_{21}, a_{41}, a_{52}, a_{13}, a_{33}, a_{24}, a_{45}),$$
$$RI = (2, \quad 4, \quad 5, \quad 1, \quad 3, \quad 2, \quad 4),$$
$$CIP = (1, \quad 3, \quad 4, \quad 6, \quad 7).$$

The above storage scheme is easy to use. For example, a_{33} can be recovered as follows. Since $CIP(3) = 4$, $RI(4)$ gives the row index of the first nonzero element of column 3. Then $RI(4)$, or one of the succeeding elements of RI prior to the first nonzero element of column

4, must contain the index 3 if $a_{33} \neq 0$. In this case, RI(5) $= 3$, hence VE(5) contains a_{33}.

SCHEME III

   With each nonzero element of the given matrix a unique integer $\lambda(i,j)$ is associated as follows:

$$\lambda(i,j) = i + (j-1)n, \qquad a_{ij} \neq 0.$$

The storage consists of two arrays, VE (Value of nonzero Elements) and LD (Lambda), each having τ elements. LD(α) contains the $\lambda(i,j)$ corresponding to the a_{ij} in VE(α), where $\alpha = 1, 2, \ldots, \tau$.
   The matrix A_5 is stored as

$$VE = (a_{21}, a_{41}, a_{52}, a_{13}, a_{33}, a_{24}, a_{45}),$$

$$LD = (2, \quad 4, \quad 10, \quad 11, \quad 13, \quad 17, \quad 24).$$

   The original matrix can be recovered from this storage scheme as follows: It is evident from the definition of $\lambda(i,j)$ given above that

$$j \text{ is the least integer} \geq \lambda(i,j)/n$$

and

$$i = \lambda(i,j) - (j-1)n.$$

For example, if $\lambda(i,j) = LD(5) = 13$, then $\lambda(i,j)/n = 13/5$, and the least integer greater than or equal to $\lambda(i,j)/n$ is 3, therefore $j = 3$ and $i = \lambda(i,j) - (j-1)n = 13 - 10 = 3$.

SCHEME IV

   If A is a symmetric matrix such that for all $i \geq j$,

$$a_{ij} = 0, \qquad i - j > \theta_i,$$

where θ_i is much smaller than n and is generally different for each value of i, then A is called a *symmetric band matrix* having a *locally variable bandwidth.* A detailed description of band matrices is given in Section 3.8. Only the lower triangular part of the matrix A which lies on or below the main diagonal is stored because A is symmetric. The storage consists of two arrays, VE (Value of Elements) and PD (Position of the Diagonal elements in VE). For each row, the left-most nonzero element and all the other elements to its right, up to and including the diagonal element are stored in VE. Thus the ith row of A needs $\theta_i + 1$ storage locations and VE will have $\sum_{i=1}^{n} (\theta_i + 1)$ elements; adding the n elements needed for PD to this, we conclude that a total of $\sum_{i=1}^{n} \theta_i + 2n$ locations are required to store A. If the bands are *full*, that is, $a_{ij} \neq 0$, for all $i - j \leqslant \theta_i$ with $i > j$, then $\sum_{i=1}^{n} \theta_i = (\tau - n)/2$ and the total storage is $(\tau + 3n)/2$.

We will illustrate the above storage scheme for the matrix whose non-zero elements which lie on or below the diagonal are $a_{11}, a_{21}, a_{22}, a_{32}, a_{33}, a_{42}, a_{44}, a_{52}, a_{53}, a_{55}$. We first notice that the leftmost nonzero element in the third, fourth, and fifth rows all lie in the second column. Therefore, the zero elements a_{43} and a_{54} must be stored. The storage is as follows:

$$\text{VE} = (a_{11}, a_{21}, a_{22}, a_{32}, a_{33}, a_{42}, 0, a_{44}, a_{52}, a_{53}, 0, a_{55}),$$
$$\text{PD} = (1, \quad 3, \quad 5, \quad 8, \quad 12).$$

An element a_{ij} of the original matrix can be recovered from the above storage scheme as follows. $\text{PD}(i) - (i - j)$ is the position of a_{ij} in VE provided that $\text{PD}(i) - (i - j) > \text{PD}(i - 1)$; that is, a_{ij} does not lie to the left of the first nonzero element of the ith row, in which case $a_{ij} = 0$ and is not stored in VE. For example, to recover a_{53} from VE, we have

$$\text{PD}(5) - (5 - 3) = 12 - 2 = 10 > 8 = \text{PD}(4)$$

hence a_{53} is stored in VE(10).

The main advantage of this packing scheme is as follows: If during some computational procedure (for example the Gaussian elimination: Chapter 2) additional nonzeros are created only to the right of the left-most nonzero element in each row, then they can be stored in VE without moving all the subsequent elements.

SOME REMARKS ON PACKING SCHEMES

We shall now briefly mention ways by which it is possible to make some additional savings in storage for the packing schemes described above.

If A is symmetric, then as in Scheme IV, only the nonzero elements in its lower (or upper) triangular portion along with the diagonal are stored.

It is possible to store two or more items of linked lists in blocks of consecutive locations. In this way, all items within a block, except the last one, require two instead of three storage locations. Of course, the insertion or deletion of an item wthin the block becomes rather difficult with this form of storage.

In all the storage schemes described above, if the length (number of binary bits) of each storage location of the computer is reasonably large, then two or more row (or column) indices can be stored at one location to save storage. This entails some knowledge of lower-level programming language and is therefore generally not satisfactory for higher-level languages like Fortran or Algol.

Except for Scheme IV, storage by columns has been described in this section. In many applications storage by rows is advantageous. Since it is very similar to storage by columns (it is the same as storing the transpose of A by columns), there is no need to describe it.

1.4. Scaling

The matrix A is often associated with a system of linear equations $Ax = b$, where x and b are both column vectors of order n. Often, the elements x_j and b_i of the vectors x and b, respectively, are measured in units that differ vastly in magnitude. For example, b_1 is measured in centimeters and b_2 in kilometers, with the result that the first row of A and b_1 are much larger (in some norm) than the second row and b_2. This situation can be improved by making the two rows equal in some norm. In view of this, it is generally recommended that the rows and columns of the given matrix be transformed in such a manner that they have magnitudes of similar order. This is called *scaling*.

A simple method of scaling the matrix A consists of dividing each row by the element having the largest absolute value in that row. This row scaling can be preceded by column scaling, namely, dividing each column of A by the element having the maximum absolute value in that column. Many linear programming codes allow this option because the computed inverse of a matrix generally has less round-off errors if it is scaled prior to the inversion.

We can describe the effect of the above row–column scaling in the solution of $Ax = b$ as follows: Let e_j denote the jth column of the nth-order identity matrix I_n. Then the solution x of $Ax = b$ is the same as that of

$$(1.4.1) \qquad D_2 A D_1 D_1^{-1} x = D_2 b,$$

where D_1 and D_2 are diagonal matrices such that

$$(1.4.2) \quad e_j' D_1 e_j = [\max_i |a_{ij}|]^{-1}; \qquad e_i' D_2 e_i = [\max_j |e_i' A D_1 e_j|]^{-1}.$$

The solution of (1.4.1) is given by

$$(1.4.3) \qquad x = D_1 (D_2 A D_1)^{-1} D_2 b,$$

thus instead of computing A^{-1}, we compute the inverse of the scaled matrix $D_2 A D_1$. If $D_1 = I_n$, then we only have *row scaling* and (1.4.3) becomes

$$(1.4.4) \qquad x = (D_2 A)^{-1} D_2 b.$$

1.5. *Bibliography and Comments*

Some references to sparse matrices in the application areas follow.

(a) Linear programming: Markowitz (1957), Larson (1962), Wolfe and Cutler (1963), Dantzig (1963a, b), Smith and Orchard-Hays (1963), Dickson (1965), Tewarson (1966, 1967a), Orchard-Hays (1968), Dantzig *et al.* (1969), Orchard-Hays (1969), Smith (1969),

Wolfe (1969), Tomlin (1970), Beale (1971), de Buchet (1971), Carré (1971), Forrest and Tomlin (1972).

(b) Structural analysis: Livesley (1960–1961), Steward (1962), Alway and Martin (1965), Jennings (1968), Rosen (1968), Cuthill and McKee (1969), McCormick (1969), Palacol (1969), Allwood (1971), Cuthill (1971), George (1972).

(c) Network theory and power distribution systems: Branin (1959), Roth (1959), Kron (1963), Sato and Tinney (1963), Tinney and Walker (1967), Edelman (1968), Chang (1969), Tinney (1969), Baty and Stewart (1971), Baumann (1971), Churchill (1971), Ogbuobiri (1971).

(d) Numerical solution of differential equations: Varga (1962), Carré (1966), Liniger and Willoughby (1969), Gear (1971), Evans (1972), Guymon and King (1972).

(e) Graph theory: Busacker and Saaty (1965), Dulmage and Mendelsohn (1967), Harary (1967, 1971a, b), Bellman *et al.* (1970).

(f) Genetic theory: Fulkerson and Gross (1965).

(g) Behavioral sciences: Harary (1960), Marimont (1959), Ross and Harary (1959).

(h) Computer programming: Marimont (1960).

(i) Other areas: Ashkenazi (1971), Glaser (1972), Guymon and King (1972).

Linked Lists are described in Maurser (1968), Kettler and Weil (1969), Ogbuobiri (1970), Churchill (1971), Gustavson (1972). The other storage schemes are given in Jennings (1966), Jimenez (1969), Jenson and Parks (1970), Nuding and Kahlert-Warmbold (1970), Berry (1971), de Buchet (1971), Gustavson (1972), Jennings and Tuff (1971). A general introduction to sparsity techniques is given by Tinney and Ogbuobiri (1970).

The row scaling (even preceded by column scaling) does not yield an optimum solution to the scaling problem. The reader is referred to Wilkinson (1965), Forsythe and Moler (1967), and Westlake (1968) for a further discussion of scaling, which is sometimes called *equilibration.*

More advanced scaling methods are discussed in Fulkerson and Wolfe (1962), Bauer (1963), van der Sluis (1969), and Curtis and Reid (1971b).

In this book we shall be mainly concerned with direct methods for inverting sparse matrices. We will not describe the indirect (or iterative) methods for sparse matrices, as they are well known (Varga, 1962). Direct methods for computing the inverse have usually been considered for small full matrices in general numerical analysis books and periodicals. But in many areas of application (for example, linear programming, structural analysis, electrical networks, and power generation and distribution systems), direct methods for inverting large sparse matrices have been extensively developed and incorporated into computer programs: Livesley (1960–1961), Smith and Orchard-Hays (1963), Dickson (1965), Jennings (1968), Dantzig *et al.* (1969), Lee (1969), Gustavson *et al.* (1970), Berry (1971), de Buchet (1971), Cantin (1971), Forrest and Tomlin (1972).

CHAPTER

2

The Gaussian Elimination

2.1. Introduction

In this chapter we describe the Gaussian elimination method for the solution of a system of linear equations. We show how the matrices associated with the various stages of the elimination process can be utilized to express the inverse of the coefficient matrix of the linear equations in a factored form. Some theorems are proved and then they are used to determine sparse factored forms of inverses of sparse matrices.

2.2. The Basic Method

The best known method for solving the system of equations

(2.2.1) $$Ax = b$$

15

(where as in Chapter 1, x and b are both column vectors of order n and A is a nonsingular matrix of order n) is the Gaussian elimination method (Wilkinson, 1965). It consists of two parts, the *forward course* in which a sequence of elementary transformations (row operations) is applied to A to reduce it to a unit upper triangular matrix U and the so-called *back substitution* in which U is inverted.

The forward course of the Gaussian elimination consists of n steps. Let $A^{(k)}$ denote the matrix at the beginning of the kth step, where $A^{(1)} \equiv A$ and $A^{(n+1)} \equiv U$. Let $a_{ij}^{(k)}$ denote the ith row and the jth column element [the (i, j) element] of $A^{(k)}$; in other words, $a_{ij}^{(k)} = e_i' A^{(k)} e_j$, where we recall from Section 1.3 that e_i is the ith column of the nth order identity matrix I_n. The matrix $A^{(k)}$ is already in upper triangular form for the first $k - 1$ columns (see Fig. 2.2.1). At the kth step, the kth row of

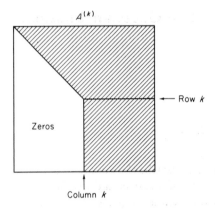

Fig. 2.2.1. The matrix at the beginning of the kth step.

$A^{(k)}$ is divided by its (k, k) element and its multiples are subtracted from all the following rows such that all the nonzero elements in the kth column that lie below the kth row become zero. The resulting matrix is denoted by $A^{(k+1)}$. This process can be restated in matrix notation as follows.

The forward course of the Gaussian elimination consists of computing

$$(2.2.2) \qquad A^{(k+1)} = L_k A^{(k)}, \qquad k = 1, 2, \ldots, n,$$

where the elementary lower triangular matrix L_k (see Fig. 2.2.2) is given

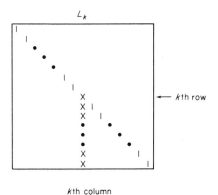

Fig. 2.2.2. The elementary matrix at the kth stage.

by

(2.2.3) $$L_k = I_n + (\eta^{(k)} - e_k)e_k',$$

with the elements of the column vector $\eta^{(k)}$ defined as follows,

(2.2.4)
$$\eta_i^{(k)} = 0, \quad i < k;$$
$$\eta_k^{(k)} = 1/a_{kk}^{(k)}, \qquad \eta_i^{(k)} = -a_{ik}^{(k)}/a_{kk}^{(k)}, \qquad i > k.$$

Thus L_k consists of ones on the diagonal except in the kth column which has $\eta_i^{(k)}$s on and below the diagonal. All other elements of L_k are zero.

Now, from (2.2.2), we have

(2.2.5) $$A^{(n+1)} = L_n \cdots L_2 L_1 A^{(1)},$$

and if we let

(2.2.6) $$L = L_n \cdots L_2 L_1,$$

then in view of the facts that $A^{(1)} \equiv A$ and $A^{(n+1)} \equiv U$, equations (2.2.5) and (2.2.6) give

(2.2.7) $$U = LA.$$

Thus the forward course of the Gaussian elimination consists of finding a lower triangular matrix L (the product of lower triangular matrices is lower triangular) that transforms A to an upper triangular matrix U. In view of (2.2.7) and the fact that the L_ks are applied to both sides of (2.2.1), it follows that at the conclusion of the Gaussian elimination, we get

(2.2.8) $Ux = Lb.$

The back substitution part of the Gaussian elimination consists of solving (2.2.8) as follows: Let x_i denote the ith element of x. Then the last element x_n is equal to the last element of the column vector Lb, since the last row of U consists of all zeros except a one for the last element. This value for x_n is substituted in the preceding equation and x_{n-1} is easily computed. x_n and x_{n-1}, when substituted in the $(n-2)$th row of (2.2.8), yield x_{n-2} and so on.

In order to express the back substitution described above in matrix notation, first we note that the (i, j) element of U is $a_{ij}^{(i+1)}$. This follows from the fact that in (2.2.2) the ith row of A is modified until $k = i$ and then remains unchanged afterwards; in other words, the ith rows of both $A^{(i+1)}$ and U are identical.

The back substitution part of the Gaussian elimination can now be defined as follows:

(2.2.9) $U_2 \cdots U_{n-1} U_n U = I_n,$

where

(2.2.10) $U_k = I_n + \xi^{(k)} e_k', \qquad k = n, n-1, \ldots, 2,$

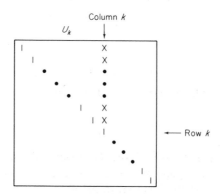

Fig. 2.2.3.   The elementary matrix at the kth stage of back substitution.

and the elements of column vector $\xi^{(k)}$ are given by

(2.2.11) $\xi_i^{(k)} = -a_{ik}^{(i+1)}, \quad i < k \qquad$ and $\qquad \xi_i^{(k)} = 0, \quad i \geqslant k$

(see Fig. 2.2.3). Thus U_k has all ones on the diagonal and $\xi_i^{(k)}$'s above the diagonal in the kth column. All other elements of U_k are zeros.

The total effect of the forward course and the back substitution on A can now be described. From (2.2.9), we get

(2.2.12) $U^{-1} = U_2 \cdots U_{n-1} U_n,$

and from (2.2.8), (2.2.6), and (2.2.12), it follows that

(2.2.13) $x = U^{-1}Lb$

$$= U_2 \cdots U_{n-1} U_n L_n \cdots L_2 L_1 b.$$

2.3.   Pivoting and Round-off Errors

The element $a_{kk}^{(k)}$ in (2.2.4) is called the *pivot* at the kth stage of elimination. Since the computations are performed by using finite-length storage locations for numbers, *round-off* errors are generally introduced. In order to minimize the effect of round-off errors in the Gaussian elimination, Wilkinson (1965) recommends the following techniques for full (nonsparse) matrices. His recommendations are based on the fact that error bounds can be obtained in such cases, and the computations are shown to be stable.

The first recommendation, called *partial pivoting* is as follows: At the kth stage, choose the element having the maximum absolute value in the kth column of $A^{(k)}$ that lies on or below the kth row, namely,

(2.3.1) $|a_{sk}^{(k)}| = \max_i |a_{ik}^{(k)}|, \qquad k \leqslant i \leqslant n,$

and interchange the sth and the kth rows of $A^{(k)}$ before the kth step in (2.2.2). Of course, all such row permutations have to be saved for later use.

The second technique for minimizing the effect of round-off errors, called *complete pivoting*, can be described as follows. At each stage k, the element having the largest absolute value is chosen from the last

$n - k + 1$ rows and columns of $A^{(k)}$, namely,

$$(2.3.2) \qquad |a_{st}^{(k)}| = \max_{i,j} |a_{ij}^{(k)}|, \qquad k \leqslant i, j \leqslant n,$$

and rows s and k and columns t and k are interchanged prior to the kth step of (2.2.2). These row and column permutations are saved for later use in unscrambling the solution.

In many practical applications involving large sparse matrices, in place of partial or complete pivoting, it generally suffices to make sure that all pivots are greater than a certain chosen number ε, called the *pivot tolerance* (Tewarson, 1969a). This is especially true in linear programming (Clasen, 1966) where the matrices are generally large and sparse. The pivot tolerance ε avoids the choice of small pivots which lead to round-off difficulties. In practice, for most problems that involve large sparse matrices $\varepsilon = 10^{-3}$ has been found satisfactory if 9 or 10 decimal digits are used for storing the nonzero elements during the computation. Of course, for full (nonsparse) matrices partial or complete pivoting should be used. It is recommended that if some elements become very small in the course of computations (less than the so-called *drop tolerance*), then they should be set to zero (Clasen, 1966). A drop tolerance of 10^{-7} has been suggested by Wolfe (1965).

2.4. The Elimination Form of Inverse

A factored form of A^{-1} can be obtained from Section 2.2 as follows: The usual solution of (2.2.1) for arbitrary b is given by $x = A^{-1}b$. Therefore, comparing this solution with the one given by (2.2.13), we conclude that

$$(2.4.1) \qquad A^{-1} = U_2 \cdots U_{n-1} U_n L_n \cdots L_2 L_1,$$

which is called the *Elimination Form of Inverse* (*EFI*). Thus A^{-1} is expressed as a product of n lower and $n - 1$ upper triangular matrices.

One of the principal advantages of EFI is the ease with which it can be post- or premultiplied by a given vector as shown below.

Let ρ_i denote the ith element of a column vector ρ. Then from (2.2.10) and (2.2.11) it follows that (see also Fig. 2.2.3).

(2.4.2)
$$e_i'(U_k\rho) = \rho_i + \xi_i^{(k)}\rho_k, \qquad i < k;$$
$$e_i'(U_k\rho) = \rho_i, \qquad i \geqslant k.$$

Thus the premultiplication of a column vector by U_k is equivalent to adding to the given column vector its kth element times $\xi^{(k)}$. We also have

(2.4.3) $(\rho'U_k)e_i = \rho_i, \qquad i \neq k; \qquad (\rho'U_k)e_k = \rho'\xi^{(k)} + \rho_k,$

and therefore the postmultiplication of the row vector ρ' by U_k leaves all elements of ρ' the same except the kth element, to which the inner product of ρ' and $\xi^{(k)}$ is added. From (2.2.3) and (2.2.4) we have (see also Fig. 2.2.2)

(2.4.4)
$$e_i'(L_k\rho) = \rho_i, \qquad i < k;$$
$$e_k'(L_k\rho) = \eta_k^{(k)}\rho_k,$$
$$e_i'(L_k\rho) = \eta_i^{(k)}\rho_k + \rho_i, \qquad i > k.$$

Thus $L_k\rho$ can be obtained from ρ by replacing its kth element by zero and then adding $\rho_k\eta^{(k)}$ to it. We also have

(2.4.5) $(\rho'L_k)e_i = \rho_i', \quad i \neq k; \qquad (\rho'L_k)e_k = \rho'\eta^{(k)},$

therefore postmultiplication of a row vector by L_k leaves all its elements the same except the kth element which is replaced by the inner product of the vector and $\eta^{(k)}$. The computation of $A^{-1}\pi$, where π is a column vector and A^{-1} is in EFI given by (2.4.1), can be arranged such that for the first n steps (2.4.4) and for the last $n - 1$ steps (2.4.2) is used. The column vector formed at each step is used in the next step as follows:

$$A^{-1}\pi = U_2(\cdots(U_n(L_n\cdots(L_2(L_1\pi))\cdots).$$

Similarly, πA^{-1} can be computed by using (2.4.3) and (2.4.5). The row vector formed at each step is used in the next step as follows:

$$\pi A^{-1} = (\cdots((\pi U_2)U_3)\cdots U_n)L_n)\cdots L_1).$$

Now, in view of (2.2.3), (2.2.4), (2.2.10), (2.2.11), and (2.4.1), it is evident that, in order to compute the EFI, we only need the nonzero elements of the $\eta^{(k)}$s and the $\xi^{(k)}$s. Therefore, only these elements (along with the relevant bookkeeping information) are stored. The computer storage is at a premium for a large sparse matrix and therefore it is important that its EFI be determined such that the storage is minimized. In other words, the number of nonzero elements in all the $\eta^{(k)}$s and $\xi^{(k)}$s should be minimized. We will now show how this can be done.

2.5.   Minimizing the Total Number of Nonzero Elements in EFI

We recall from Section 2.2 that at each stage of the Gaussian elimination multiples of a row are subtracted from several other rows of the matrix. This generally leads to the creation of new nonzero elements in place of zero elements. For example, if at the kth stage, $a_{kk}^{(k)}$, $a_{kj}^{(k)}$, and $a_{ik}^{(k)}$ are all nonzero but $a_{ij}^{(k)} = 0$, where $i, j > k$, then from (2.2.2), (2.2.3), and (2.2.4) it follows that at the end of the kth stage

$$(2.5.1) \qquad\qquad a_{ij}^{(k+1)} = -(a_{ik}^{(k)}/a_{kk}^{(k)})a_{kj}^{(k)},$$

which is clearly nonzero. Thus, under the above-mentioned conditions a zero in (i, j) position of $A^{(k)}$ has become a nonzero in $A^{(k+1)}$. The total number of all such elements that change from zero in $A^{(k)}$ to nonzero in $A^{(k+1)}$ is called the local *fill-in*.

If instead of choosing a $a_{kk}^{(k)}$ as the pivot at the kth stage, we select another nonzero element $a_{st}^{(k)}$, $s \geqslant k, t \geqslant k$ as the pivot, then we have to interchange the kth and the sth rows, as well as the kth and the tth columns of $A^{(k)}$ before computing $A^{(k+1)}$ in (2.2.2). Thus instead of (2.2.2), we have

$$(2.5.2) \qquad\qquad A^{(k+1)} = L_k \hat{A}^{(k)}, \qquad k = 1, 2, \ldots, n$$

where

$$(2.5.3) \qquad\qquad \hat{A}^{(k)} = P_k A^{(k)} Q_k,$$

and both P_k and Q_k are the matrices obtained from the identity matrix I_n by interchanging its kth and sth rows and the kth and tth columns, respectively. The L_ks are the same as in (2.2.3), but the $\eta^{(k)}$s are given by

(2.5.4)
$$\eta_i^{(k)} = 0, \quad i < k;$$
$$\eta_k^{(k)} = 1/\hat{a}_{kk}^{(k)}, \qquad \eta_i^{(k)} = -\hat{a}_{ik}^{(k)}/\hat{a}_{kk}^{(k)}, \qquad i > k,$$

where $\hat{a}_{ij}^{(k)}$ is the (i, j) element of $\hat{A}^{(k)}$.

In view of the last paragraph of Section 2.3, we recall that $|a_{st}^{(k)}| > \varepsilon$, where ε is the pivot tolerance. Therefore, from all the available candidates for pivot, namely, $|a_{st}^{(k)}| > \varepsilon$ with $i \geqslant k, j \geqslant k$, the pivot chosen is the one that leads to least local fill-in. This can be done as follows.

> *DEFINITION*   Let B_k be the matrix obtained from the last $n - k + 1$ rows and columns of $A^{(k)}$ by replacing all the nonzero elements by unity.

The following theorem (Tewarson, 1967b) can be used to determine the pivot which leads to the least fill-in.

(2.5.5)   *THEOREM*   If $a_{i+k-1,j+k-1}^{(k)}$ is chosen as a pivot at the kth stage of the forward course of the Gaussian elimination, then the local fill-in is given by the (i, j) element of the matrix G_k, with

(2.5.6)
$$G_k = B_k \bar{B}_k' B_k,$$

where $\bar{B}_k'$ is the transpose of the matrix which results when each zero element of B_k is changed to unity and vice-versa.

Proof   If in $A^{(k)}$, the $(p + k - 1, q + k - 1)$ element is zero but both $(i + k - 1, q + k - 1)$ and $(p + k - 1, j + k - 1)$ are nonzero, then from (2.5.2), (2.2.3), (2.5.3), and (2.5.4) it follows that the $(p + k - 1, q + k - 1)$ element in $A^{(k+1)}$ will be nonzero. This is equivalent to saying that if

$$b_{pq}^{(k)} = 0 \qquad \text{and} \qquad b_{iq}^{(k)} = b_{pj}^{(k)} = 1,$$

where $b_{pq}^{(k)}$ is the (p, q) element of B_k, then one new nonzero element is created. If $g_{ij}^{(k)}$ denotes the total number of such new nonzero elements created (the local fill-in) at the kth stage of the Gaussian elimination, then

(2.5.7)
$$g_{ij}^{(k)} = \sum_p \sum_q b_{iq}^{(k)}(1 - b_{pq}^{(k)})b_{pj}^{(k)}, \qquad p \neq i, \quad q \neq j$$

$$= \sum_p \sum_q e_i' B_k e_q (1 - e_p' B_k e_q) e_p' B_k e_j,$$

where we have dropped the restriction $p \neq i$, $q \neq j$, since for $p = i$ either $b_{iq}^{(k)} = 0$ or $1 - b_{pq}^{(k)} = 0$, and for $q = j$ either $1 - b_{pq}^{(k)} = 0$ or $b_{pj}^{(k)} = 0$. Now if M is an $(n - k + 1)$th order matrix of all ones, then

(2.5.8)
$$1 - e_p' B_k e_q = e_p' M e_q - e_p' B_k e_q = e_p'(M - B_k)e_q$$

$$= e_p' \bar{B}_k e_q = e_q' \bar{B}_k' e_p,$$

where we have used the fact that $e_p' \bar{B}_k e_q$ is a scalar quantity. From (2.5.7) and (2.5.8), we have

$$g_{ij}^{(k)} = \sum_p \sum_q e_i' B_k e_q e_q' \bar{B}_k' e_p e_p' B_k e_j$$

$$= e_i' B_k \sum_q e_q e_q' \bar{B}_k' \sum_p e_p e_p' B_k e_j$$

$$= e_i' B_k \bar{B}_k' B_k e_j,$$

since $\sum_q e_q e_q' = \sum_p e_p e_p' = I_{n-k+1}$. This completes the proof of the theorem.

Now, we are finally in a position to choose a pivot which leads to least local fill-in. This can be done by using the following corollary, the proof of which is omitted as it is a direct consequence of Theorem 2.5.5.

(2.5.9)   COROLLARY   If at the kth stage of the Gaussian elimination $a_{st}^{(k)}$ is chosen as the pivot, where $s = \alpha + k - 1$, $t = \beta + k - 1$ and α, β are given by

(2.5.10)
$$g_{\alpha\beta}^{(k)} = \min_{i,j} e_i' G_k e_j \qquad \text{for all} \quad |a_{i+k-1, j+k-1}^{(k)}| > \varepsilon,$$

(ε is some suitably chosen pivot tolerance), then the local fill-in will be minimized.

In view of (2.2.11), the nonzero elements of $\xi_i^{(k)}$'s are obtained by changing the signs of the nonzero elements of U which lie above the diagonal; therefore from (2.2.9) and (2.2.10) it follows that the back substitution part of the Gaussian elimination does not create any new nonzero elements. Thus the new nonzeros are created only in the forward course. At the end of the kth stage, the last $n - k$ rows and columns of $A^{(k+1)}$ generally have some nonzero elements where in $A^{(k)}$ the corresponding elements were zeros. Since such rows and columns are utilized in subsequent stages for computing the $\eta^{(k)}$'s and $\xi^{(k)}$'s, minimizing the local fill-in will minimize the number of nonzero elements in $\eta^{(k)}$'s and $\xi^{(k)}$'s, provided that local minima lead to a global minimum. This may be true for some matrices but is not so for arbitrary sparse matrices. In any case, minimizing the local growth of such nonzero elements by making use of Corollary 2.5.9, still leads to a substantial decrease in the nonzero elements of all the $\xi^{(k)}$'s and the $\eta^{(k)}$'s.

The choice of the minimum fill-in pivot $a_{st}^{(k)}$ entails no special problems. The relevant changes can be described as follows. From (2.5.2) and (2.5.3), we have

(2.5.11) $A^{(n+1)} = L_n P_n \cdots L_2 P_2 L_1 P_1 A Q_1 Q_2 \cdots Q_n,$

and if we let

(2.5.12) $\hat{L} = L_n P_n \cdots L_2 P_2 L_1 P_1,$

$$Q_1 Q_2 \cdots Q_n = Q \quad \text{and} \quad A^{(n+1)} = \hat{U},$$

then we get

(2.5.13) $A^{-1} = Q \hat{U}^{-1} \hat{L}.$

The permutation matrices Q and P_ks require a storage of order n (and not n^2), since in each case only the positions of the nontrivial elements need to be stored.

A simpler, though less accurate, way of finding a pivot which tends to keep the local fill-in small is based on the following theorem (Markowitz, 1957).

(2.5.14)   *THEOREM*   If $a_{i+k-1, j+k-1}^{(k)}$ is chosen as a pivot at the kth stage of the Gaussian elimination, then the maximum possible fill-in (not the actual fill-in) is given by the (i, j) element of $\hat{G}_k$, with

(2.5.15) $$\hat{G}_k = (B_k - I_{n-k+1})M(B_k - I_{n-k+1}),$$

where M is a matrix of all ones.

Proof If $\hat{g}_{ij}^{(k)}$ denotes the maximum possible fill-in at the kth stage, then as in the proof of Theorem 2.5.5, we have

(2.5.16) $$\hat{g}_{ij}^{(k)} = \sum_p \sum_q b_{iq}^{(k)} b_{pj}^{(k)}, \qquad \text{where } p \neq i \quad \text{and} \quad q \neq j,$$

$$= \left(\sum_q b_{iq}^{(k)} - b_{ij}^{(k)} \right) \left(\sum_p b_{pj}^{(k)} - b_{ij}^{(k)} \right)$$

$$= \left(\sum_q b_{iq}^{(k)} - 1 \right) \left(\sum_p b_{pj}^{(k)} - 1 \right), \qquad \text{since } b_{ij}^{(k)} = 1$$

$$= \left(e_i' B_k \sum_q e_q - 1 \right) \left(\sum_p e_p' B_k e_j - 1 \right)$$

$$= (e_i' B_k V_k - e_i' V_k)(V_k' B_k e_j - V_k' e_j),$$

where V_k is a column vector of all ones of order $n - k + 1$. Thus

$$\hat{g}_{ij}^k = e_i'(B_k - I_{n-k+1})V_k V_k'(B_k - I_{n-k+1})e_j$$

$$= e_i'(B_k - I_{n-k+1})M(B_k - I_{n-k+1})e_j,$$

since $V_k V_k' = M$. This completes the proof of the theorem.

To make use of the above theorem, we use the following equation [instead of (2.5.10)] to select the pivot at the kth stage.

(2.5.17) $$\hat{g}_{\alpha\beta}^{(k)} = \min_{i,j} \hat{g}_{ij}^{(k)} \qquad \text{for all} \quad |a_{i+k-1, j+k-1}^{(k)}| > \varepsilon,$$

where, as before, $\alpha + k - 1 = s$ and $\beta + k - 1 = t$. We note that the pivot $a_{st}^{(k)}$ chosen according to (2.5.17) does not necessarily lead to the least local fill-in.

An interesting application of the above pivot selection scheme is when $B_k = B_k'$ and only the diagonal elements are chosen as pivots.

In this case, from (2.5.16), it follows that

$$\hat{g}_{ii}^{(k)} = (e_i'B_kV_k - 1)(V_k'B_ke_i - 1)$$
$$= (e_i'B_kV_k - 1)^2, \quad \text{since} \quad B_k' = B_k.$$

Therefore

$$\hat{g}_{\alpha\alpha}^{(k)} = \min_i (e_i'B_kV_k - 1)^2,$$

but the index α is the same for $\min_i (e_i'B_kV_k - 1)$ or $\min_i e_i'B_kV_k$, since $e_i'B_kV_k \geqslant 1$. In other words,

$$(2.5.18) \qquad \min_i e_i'B_kV_k = e_\alpha'B_kV_k,$$

with

$$|a_{\alpha+k-1,\alpha+k-1}^{(k)}| > \varepsilon,$$

is used for selecting the pivot from the diagonal elements. Note that $e_i'B_kV_k$ is the total number of nonzero elements in the $(i + k - 1)$th row of $A^{(k)}$. Thus the row (and the corresponding column) having the least number of nonzero elements is selected at each stage as pivot row (column). This is relatively simple and easy to do, and is therefore recommended in many practical applications (Tinney and Walker, 1967; Spillers and Hickerson, 1968; Churchill, 1971). One of the main reasons for choosing only the diagonal elements as pivots is that if A is symmetric, then quite often only the upper triangular part of A along with the main diagonal is stored and the diagonal pivot choice during the forward course of the Gaussian elimination maintains the symmetry. Furthermore the $\eta^{(k)}$s can be easily obtained from the upper triangular matrix obtained at the end of the forward course. We state these facts as a theorem.

(2.5.19)   *THEOREM*   If A is symmetric and only the diagonal elements are chosen as pivots, then (a) for $k = 1, 2, \ldots, n - 1$, the matrix consisting of the last $n - k$ rows and columns of $A^{(k+1)}$ in (2.5.2) is also symmetric; and (b), in Equation (2.5.4), $\eta_i^{(k)}, i > k$ are given by

$$(2.5.20) \qquad \eta_i^{(k)} = -a_{ki}^{(k+1)} = \xi_k^{(i)}, \qquad i > k.$$

Proof   If we can show that $a_{ij}^{(k+1)} = a_{ji}^{(k+1)}$ for $i, j > k$ whenever $a_{ij}^{(k)} = a_{ji}^{(k)}$, $i, j \geq k$, then by induction on k and the fact
that $a_{ij}^{(1)} = a_{ji}^{(1)}$, part (a) of the theorem will clearly follow.
As only the diagonal elements are chosen as pivots, therefore
in (2.5.3), $Q_k = P_k'$ and $\hat{a}_{ij}^{(k)} = \hat{a}_{ji}^{(k)}$, $i, j \geq k$. Now, from (2.5.2),
(2.2.3), and (2.5.4), for $i, j > k$, we have

$$a_{ij}^{(k+1)} = \hat{a}_{ij}^{(k)} - \hat{a}_{ik}^{(k)}\hat{a}_{kj}^{(k)}/\hat{a}_{kk}^{(k)},$$

$$a_{ji}^{(k+1)} = \hat{a}_{ji}^{(k)} - \hat{a}_{jk}^{(k)}\hat{a}_{ki}^{(k)}/\hat{a}_{kk}^{(k)},$$

and therefore it follows that

$$a_{ij}^{(k+1)} = a_{ji}^{(k+1)}, \qquad i, j > k,$$

since

$$\hat{a}_{ij}^{(k)} = \hat{a}_{ji}^{(k)}, \qquad i, j \geq k.$$

This completes the proof of (a).

Now, from (2.5.2), (2.2.3), and (2.5.4), we have for $i > k$,
$a_{ki}^{(k+1)} = \hat{a}_{ki}^{(k)}/\hat{a}_{kk}^{(k)}$ and $\eta_i^{(k)} = -\hat{a}_{ik}^{(k)}/\hat{a}_{kk}^{(k)}$, and in view of (2.2.11)
and the fact that $\hat{a}_{ik}^{(k)} = \hat{a}_{ki}^{(k)}$, we have (2.5.20); which completes
the proof of the theorem.

We conclude this section with a few comments.

If in (2.5.10) the minimum is attained for more than one pair (i, j),
then the ties can be broken by choosing the pair for which $\hat{g}_{ij}^{(k)}$ given
by (2.5.16) is a maximum. This will remove the maximum number of
nonzero elements from consideration in the next step.

Instead of (2.5.15), sometimes the matrix

(2.5.21) $$\tilde{G}_k = B_k M_k B_k$$

is used to select the pivot for the kth stage of the Gaussian elimination.
We will now show that if in the kth stage, $a_{i+k-1, j+k-1}^{(k)}$ is chosen as
the pivot then $e_i'\tilde{G}_k e_j$ is the total number of multiplications and
divisions. It is evident from the proof of Theorem 2.5.14 and equations
(2.5.2), (2.5.3), and (2.5.4) that one division is required to compute
$1/a_{i+k-1, j+k-1}^{(k)}$, and $V_k'B_k e_j - 1$ and $e_k'B_k V_k - 1$ multiplications are
needed, respectively, to compute $\eta^{(k)}$ and $e_k'A^{(k+1)}$. Furthermore, to
eliminate $a_{p+k-1, j+k-1}^{(k)} \neq 0$, $p \neq i$, a total of $(e_i'B_k V_k - 1)(V_k'B_k e_j - 1)$

multiplications are required. Thus, at the kth stage of the Gaussian elimination, the total number of multiplications and division is equal to

$$1 + (V_k'B_k e_j - 1) + (e_i'B_k V_k - 1) + (e_i'B_k V_k - 1)(V_k'B_k e_j - 1)$$

$$= e_i'B_k V_k V_k'B_k e_j = e_i'B_k M B_k e_j = e_i'\tilde{G}_k e_j.$$

In view of the above facts, if instead of (2.5.17), we use

(2.5.22) $\quad \tilde{g}_{\alpha\beta}^{(k)} = \min_{i,j} e_i'\tilde{G}_k e_j \qquad$ for all $\quad |a_{i+k-1,j+k-1}^{(k)}| > \varepsilon,$

to select the pivot at the kth stage then the total number of multiplications and divisions is minimized.

Let the total number of multiplications and divisions at each stage k be taken as the measure of the computational effort for that stage. Then to minimize both, the fill-in and the computational effort a weighted average of G_k and $\tilde{G}_k$ should be used to determine the pivot. The weighing factors determine the relative importance of the two criteria. From (2.5.21) and (2.5.6) it is evident that the matrix $B_k(M - \delta B_k')B_k$ is the weighted average of G_k and $\tilde{G}_k$, where $0 \leqslant \delta \leqslant 1$, and δ and $1 - \delta$ are, respectively, the weighing factors. The value of δ depends on the computer hardware as well as the software. It should be noted that for large sparse matrices, minimizing the local fill-in is more important than minimizing the local computational effort, because the former tends to minimize the computational effort in the later stages by keeping B_k sparse.

If the solution of (2.2.1) is needed for only a few right-hand sides, then L_ks defined by (2.2.3) and (2.5.4) are not saved (the right-hand sides are transformed at each stage). Now, in view of the fact that the last $n - k$ elements of the kth column of $\hat{A}^{(k)}$ are deleted in the kth stage, the increase in the number of nonzeros in $A^{(k+1)}$ over $\hat{A}^{(k)}$ is equal to

$$g_{ij}^{(k)} - (V_k'B_k e_j - 1) = e_i'B_k \bar{B}_k'B_k e_j - e_i'M B_k e_j + 1$$

$$= e_i'(B_k \bar{B}_k' - M)B_k e_j + 1,$$

and the minimum value of this can be used to select a pivot.

2.6. Storage and Use of the Elimination Form of Inverse

The $\eta^{(k)}$s needed in the elimination form of inverse (EFI) are stored as follows: At the kth stage of the forward course of the Gaussian elimination all $\hat{a}_{ik}^{(k)} \neq 0$, $i > k$ are transformed to zero and $\hat{a}_{kk}^{(k)}$ is transformed to unity; that is, $a_{ik}^{(k+1)} = 0, i > k$, and $a_{kk}^{(k+1)} = 1$. Therefore, from (2.5.4), it is clear that each $\eta_i^{(k)} \neq 0$, $i > k$ can be stored in place of the corresponding $\hat{a}_{ik}^{(k)} \neq 0$, $i > k$; also $\eta_k^{(k)}$ can be stored in place of $a_{kk}^{(k+1)}$, as there is no need to store $a_{kk}^{(k+1)} = 1$ (this holds for all k; in other words, the diagonal elements of U are all equal to one).

The permutation matrices P_k and Q_k in (2.5.3) can be easily constructed if s and t are known. Therefore, for each k only two storage locations are needed for storing the corresponding P_k and Q_k, which make a total of $2n$ locations for all the P_ks and Q_ks needed in (2.5.12).

We see from (2.2.12), (2.2.10), and (2.2.11) that only the nonzero elements of $\xi^{(k)}$s are needed to compute U_ks and furthermore that such elements can be obtained by changing the signs of those nonzero elements of U which lie above the diagonal. Therefore, the nonzero $\xi^{(k)}$s can be stored in the space occupied by the nonzero elements of U.

The nonzero elements of each $A^{(k)}$ and the corresponding $\eta^{(k)}$ and $\xi^{(k)}$ are of course stored in one of the packed forms described in Section 1.3. At each stage, k additional nonzero elements are created in $A^{(k)}$ and the linked list storage is especially suitable for storing such elements (Ogbuobiri, 1970).

We conclude this section by giving some instances in which the extra work involved in getting a sparse EFI is justified.

In many practical applications, equation (2.1.1) has to be solved repeatedly for several right-hand sides and/or coefficient matrices with the same sparseness structure. For example, the standard Newton's method for solving nonlinear equations leads to (2.2.1), where A has a fixed sparseness structure and b changes from case to case (Liniger and Willoughby, 1969; Churchill, 1971). In structural analysis the solution of (2.2.1) is required for many right-hand sides (Allwood, 1971). EFI is most suited for the above-mentioned cases. Therefore, in Section 2.5, the cost of analysis for minimizing the total number of nonzero elements in EFI can be written off over the repeated solutions of (2.2.1). The use of EFI in the solution of power system problems has in practice led to gains in speed, storage, and accuracy that are approximately proportional to the degree of sparsity (Tinney, 1969).

2.7. Bibliography and Comments

The basic Gaussian elimination method and round-off errors are discussed in Fox (1965), Wilkinson (1965), and Forsythe and Moler (1967).

The EFI and techniques for keeping it sparse by minimizing the local fill-in have been the subject of extensive attention, for example, Markowitz (1957), Dantzig (1963b), Carpentier (1963), Edelman (1963), Sato and Tinney (1963), Tewarson (1967b), Spillers and Hickerson (1968), Brayton *et al.* (1969), Ogbuobiri (1970), Tomlin (1970), Berry (1971), Bertelé and Brioschi (1971), Forrest and Tomlin (1972), and several articles in Willoughby (1969) and Reid (1971).

The EFI of A was first suggested by Markowitz (1957), and later on by Dantzig (1963b). In many applications the techniques for minimizing the fill-in do not create any round-off problems: for example, Churchill (1971) observes that in solving complex power flow problems no round-off problems were encountered when the methods for minimizing the fill-in were used. Carré (1971) has shown that the techniques for minimizing the fill-in can also be used in minimal cost network flow problems.

In Chapter 3 we shall discuss additional methods for creating sparse EFI. Some of these methods will involve the a priori row-column permutation of A to a form in which the fill-in is limited to only some areas of A.

Additional Methods for Minimizing the Storage for EFI

3.1. Introduction

In this chapter we shall be concerned mainly with methods which tend to keep the fill-in small during the forward course of the Gaussian elimination and at the same time do not require much work. These methods are generally labeled as "a priori" methods as the information for the pivot choice at each stage is obtained primarily from the original matrix and not from its transformed forms at each stage; this generally leads to significant savings in time and effort. However, such a priori methods are generally less efficient for minimizing the local fill-in than the methods given in Chapter 2. The methods in this chapter attempt to achieve a sparse EFI. They will be classified into two categories: first, those that primarily involve a priori arrangement of columns; and second, those consisting of a priori row and column permutations to transform A into various forms that are desirable for Gaussian elimination. In these desirable forms, during the forward course of the Gaussian elimination, either there is no fill-in or the fill-in is limited to only certain known regions of the matrix.

3.2. Methods Based on A Priori Column Permutations

Our aim is to determine a permutation matrix Q, prior to the start of the Gaussian elimination, and another permutation matrix P as the elimination proceeds, such that

$(3.2.1)$ $$PAQ = \hat{A},$$

where $\hat{A}$ is such that its diagonal elements sequentially taken as pivots, starting from the top left-hand corner, lead to the smallest amount of total fill-in. It is, of course, desirable to minimize the effort required to determine P and Q. We shall describe some a priori methods of finding an approximation for Q; these will be followed by a method, for the determination of an approximation for P, which is based partly on the information at each step of the forward course of the Gaussian elimination.

It is clear from (3.2.1) that Q determines the order in which the columns of A are considered for pivoting. Thus the determination of Q is equivalent to the a priori rearrangement of the columns of A. We shall now describe three methods for ordering the columns of A so as to keep the fill-in reasonably small.

> *DEFINITION*   Let $r_i^{(k)}$ and $c_j^{(k)}$ denote, respectively, the total number of nonzero elements in the ith row and the jth column of B_k, where B_k is as defined in Section 2.5.

From the above definition, it follows that

$(3.2.2)$ $$r_i^{(k)} = e_i' B_k V_k \quad \text{and} \quad c_j^{(k)} = V_k' B_k e_j,$$

where V_k is a column vector of all ones of order $n - k + 1$, and e_i is the ith column of the identity matrix of order $n - k + 1$. Now from (2.5.16) and (3.2.2), we have

$$(3.2.3) \qquad \hat{g}_{ij}^{(k)} = (r_i^{(k)} - 1)(c_j^{(k)} - 1),$$

therefore, for a given column j, the minimum value for $\hat{g}_{ij}^{(k)}$ is given by

$$(3.2.4) \quad \gamma_j^{(k)} = \min_i \hat{g}_{ij}^{(k)} = (c_j^{(k)} - 1) \min_i (r_i^{(k)} - 1),$$

$$\text{for all } i \text{ with } b_{ij}^{(k)} = 1,$$

$$= (c_j^{(k)} - 1)(r_\alpha^{(k)} - 1), \qquad \text{where} \quad b_{\alpha j}^{(k)} = 1.$$

In view of Theorem 2.5.14, it is evident that in the $(j + k - 1)$th column of $A^{(k)}$, out of all possible pivots, the one in the $(\alpha + k - 1)$th row leads to the least value for the maximum local fill-in. Therefore, to keep the fill-in small, the columns of A can be arranged a priori in ascending values of $\gamma_j^{(1)}$'s. Such an arrangement of the columns of A leads to progressively bad estimates for maximum local fill-ins as k increases from one to n, since for a given column, it may not be possible to pivot on the row that leads to a least value for maximum local fill-in, if such a row has already been pivoted earlier in other columns. The use of $\gamma_j^{(1)}$ for selecting a subset of the columns of A for pivoting has been recommended by Orchard-Hays (1968) for a closely related factored form of A^{-1}, which we discuss in Chapter 5; after the columns in the current subset have been pivoted, the next set of columns is selected on the basis of $\gamma_j^{(p)}$'s, where p is the total number of columns of A that have already been pivoted.

The second method for column ordering, which has been found useful in practice, is as follows (Tewarson, 1967b). In column j, there are $c_j^{(k)}$ nonzero elements and each one of them is a potential pivot and therefore, in view of (3.2.3), the average value for the maximum local fill-in (if each nonzero element of column j is as equally likely to be selected as a pivot as the others) is given by

$$(3.2.5) \qquad \lambda_j^{(k)} = \left[\sum_i (r_i^{(k)} - 1)(c_j^{(k)} - 1) \right] \bigg/ c_j^{(k)},$$

$$\text{for all } i \text{ with } b_{ij}^{(k)} = 1,$$

$$= (d_j^{(k)} - c_j^{(k)})(1 - 1/c_j^{(k)}),$$

where

$$(3.2.6) \qquad d_j^{(k)} = \sum_i r_i^{(k)}, \qquad \text{for all } i \text{ with } b_{ij}^{(k)} = 1.$$

Therefore, if the columns of A are arranged in ascending values of $\lambda_j^{(k)}$s, then evidently the fill-in is kept small.

A third method which is very simple but far less accurate in practice is to arrange the columns of A in the order of ascending values of $c_j^{(1)}$s. However, in practice the use of $\lambda_j^{(k)}$s or $\gamma_j^{(k)}$s leads to much better results with only a little increase in initial work (Tewarson, 1967b; Orchard-Hays, 1968).

It is also possible to express $\gamma_j^{(k)}$ in (3.2.4) and $\lambda_j^{(k)}$ in (3.2.5) as follows: From Theorem 2.5.14, we have

$$(3.2.7) \qquad \gamma_j^{(k)} = \min_i e_i' \hat{G}_k e_j,$$

$$\text{for all } i \text{ with } e_i' B_k e_j = 1.$$

Also, in view of (3.2.2), we have

$$\lambda_j^{(k)} = \left(\sum_i e_i' \hat{G}_k e_j \right) \Big/ V_k' B_k e_j,$$

$$\text{for all } e_i' B_k e_j = 1,$$

or

$$(3.2.8) \qquad \lambda_j^{(k)} = (e_j' B_k' \hat{G}_k e_j)/V_k' B_k e_j,$$

since $\sum_i e_i'$, for all i with $e_i' B_k e_j = 1$, is identical with

$$\sum_i e_j' B_k' e_i e_i' = e_j' B_k' \sum_i e_i e_i' = e_j' B_k' I_{n-k+1} = e_j' B_k'.$$

The rearrangement of the columns of A by using the $c_j^{(1)}$s, $\gamma_j^{(1)}$s or $\lambda_j^{(1)}$s leads to the matrix $\tilde{A}$, such that

$$(3.2.9) \qquad \tilde{A} = AQ.$$

If the pth column of A becomes the kth column in $\tilde{A}$, then in view of (3.2.9), we have

$$Ae_p = \tilde{A}e_k = AQe_k,$$

which implies that $e_p = Qe_k$, that is, the pth column of I_n is the kth column of Q. Thus the same reordering of columns which changes A to $\tilde{A}$ when applied to I_n will give Q.

Having obtained the matrix $\tilde{A}$ by using $c_j^{(1)}$s, $\gamma_j^{(1)}$s, or $\lambda_j^{(1)}$s, we then perform the forward course of the Gaussian elimination by taking the columns of $\tilde{A}$ sequentially, and in each column choosing a pivot which lies in a row having the least number of nonzero elements. Evidently, this pivot choice tends to keep the fill-in at each stage small. All of this can be described mathematically as follows. Let $A^{(1)} = \tilde{A}$ and in place of (2.5.3), $\hat{A}^{(k)}$ be defined by

$$\text{(3.2.10)} \qquad \hat{A}^{(k)} = P_k A^{(k)},$$

where P_k is obtained from I_n by interchanging its $(\alpha + k - 1)$th and kth rows, with the index α given by

$$\text{(3.2.11)} \qquad \tilde{r}_\alpha^{(k)} = \min_i \tilde{r}_i^{(k)}, \qquad \text{for all } i \text{ with } |\hat{a}_{i+k-1,k}^{(k)}| > \varepsilon,$$

where $\tilde{r}_i^{(k)}$ is an approximation for $r_i^{(k)}$ defined in (3.2.2). The pivot tolerance ε is the same as in (2.5.10).

The approximation $\tilde{r}_i^{(k)}$ for $r_i^{(k)}$ can be obtained by making use of the following theorem (Tewarson, 1966).

(3.2.12) *THEOREM* If the nonzero elements in the last $n - k + 1$ rows and columns of $\hat{A}^{(k)}$ are randomly distributed in those rows and columns, and $\hat{r}_i^{(k)}$ is the number of nonzero elements in the $(i + k - 1)$th row of $\hat{A}^{(k)}$, then for $i > 1$, $\tilde{r}_{i-1}^{(k+1)}$, the expected number of nonzeros in the corresponding row of $A^{(k+1)}$ is given by

$$\text{(3.2.13)} \qquad \tilde{r}_{i-1}^{(k+1)} = \hat{r}_i^{(k)}, \qquad \hat{a}_{i+k-1,k}^{(k)} = 0$$

$$\text{(3.2.14)} \qquad \tilde{r}_{i-1}^{(k+1)} = \hat{r}_i^{(k)} + \hat{r}_1^{(k)} - 2 - \frac{(\hat{r}_i^{(k)} - 1)(\hat{r}_1^{(k)} - 1)}{n - k}, \qquad \hat{a}_{i+k-1,k}^{(k)} \neq 0,$$

where $A^{(k+1)}$ is defined by (2.5.2), (3.2.10), and (2.5.4).

Proof Let us obtain $\hat{B}_k$ from the last $n - k + 1$ rows and columns of $\hat{A}^{(k)}$ by replacing its nonzeros by ones. From the definition of $\hat{B}_k$ and B_{k+1} it follows that the ith row of $\hat{B}_k$ corresponds to the $(i - 1)$th row of B_{k+1}, and for all i, j, the $(i + k - 1, j + k - 1)$ elements of $\hat{A}^{(k)}$ correspond to $\hat{b}_{ij}^{(k)}$, the (i, j) element of $\hat{B}_k$. If $\hat{a}_{i+k-1,k}^{(k)} = 0$, then $\hat{b}_{i1}^{(k)} = 0$ and clearly the ith rows of $\hat{A}^{(k)}$ and $A^{(k+1)}$ are identical, which implies that $r_{i-1}^{(k+1)} = \tilde{r}_{i-1}^{(k+1)} = \hat{r}_{i-1}^{(k)}$, thus (3.2.13) is proved.

In the other case, if $\hat{a}^{(k)}_{i+k-1,k} \neq 0$ then $\hat{b}^{(k)}_{i1} = 1$, and there is fill-in whenever $\hat{b}^{(k)}_{1j} = 1$ and $\hat{b}^{(k)}_{ij} = 0$. Let $\Gamma(\sigma)$ denote the probability that the event σ happens. Since the elements of $\hat{B}_k$ are randomly distributed because the corresponding elements of $A^{(k)}$ are, therefore,

(3.2.15) $\Gamma(\hat{b}^{(k)}_{1j} = 1 \text{ and } \hat{b}^{(k)}_{ij} = 0) = \Gamma(\hat{b}^{(k)}_{1j} = 1)\Gamma(\hat{b}^{(k)}_{ij} = 0)$

$$= \left(\frac{\hat{r}^{(k)}_1 - 1}{n - k}\right)\left(1 - \frac{\hat{r}^{(k)}_i - 1}{n - k}\right),$$

where we have made use of the fact that if the first column of $\hat{B}_k$ is excluded, then in the remaining $n - k$ columns, the proportions of nonzeros to zeros in the first and the ith row are, respectively, $(\hat{r}^{(k)}_1 - 1)/(n - k)$ and $(\hat{r}^{(k)}_i - 1)/(n - k)$, since $\hat{b}^{(k)}_{11} = \hat{b}^{(k)}_{i1} = 1$. Now, from (3.2.15), it follows that the expected value of fill-in in the ith row of $\hat{B}_k$ is equal to

$$(n - k)\left(\frac{\hat{r}^{(k)}_1 - 1}{n - k}\right)\left(1 - \frac{\hat{r}^{(k)}_i - 1}{n - k}\right),$$

adding this to $\hat{r}^{(k)}_i$ (the initial number of nonzeros) and subtracting one, since $a^{(k+1)}_{i+k-1,k} = 0$, we have

$$\tilde{r}^{(k+1)}_{i-1} = \hat{r}^{(k)}_i + (\hat{r}^{(k)}_1 - 1)\left(1 - \frac{\hat{r}^{(k)}_i - 1}{n - k}\right) - 1,$$

which on simplification yields (3.2.14). This completes the proof of the theorem.

In order to make use of the preceding theorem, we start with $\tilde{r}^{(1)}_i = r^{(1)}_i$, where $r^{(1)}_i = e'_i B_1 V_1$ and B_1 is obtained from $A^{(1)} = \tilde{A}$ [see (3.2.2) and (3.2.9)]. For each k, in view of (3.2.10) and (3.2.11) and to keep the notation simple, we use $\hat{r}^{(k)}_i$ to denote its approximate value also; in other words

(3.2.16) $\hat{r}^{(k)}_i = P_k \tilde{r}^{(k)}_i.$

Then we use (3.2.13) and (3.2.14) to update $\hat{r}^{(k)}_i$ to $\tilde{r}^{(k+1)}_{i-1}$ for all $1 < i \leqslant n - k + 1$ at each stage k. This method avoids the computation of the exact value $r^{(k)}_i$ for each k from the corresponding B_k. For large sparse matrices stored in packed form this can lead to a significant saving of time and effort. However, the $\tilde{r}_i^{(k)}$'s given by (3.2.13) and (3.2.14) were obtained under a probabilistic hypothesis, and therefore are only approximations to the actual $r_i^{(k)}$'s. For large sparse matrices and small values of k, they are reasonably good approximations, but get progressively worse as k increases. Therefore, it is recommended that, if possible, the exact values for $r_i^{(k)}$'s should be determined from the current B_k at periodic intervals.

In certain cases it is possible to determine the $r_i^{(k)}$'s exactly, without too much extra work. For example, if $A^{(k)}$ is stored in linked list form of Section 1.3 then in (2.5.2) each time a new nonzero element is created or a nonzero becomes a zero, the corresponding $r_i^{(k)}$, can be easily modified.

The methods of this section can be summarized as follows: First, we use $c_j^{(k)}$'s, $\gamma_j^{(k)}$'s, or $\lambda_j^{(k)}$'s defined in (3.2.2), (3.2.4), and (3.2.5), respectively, to determine $\tilde{A}$ and Q in (3.2.9). Then we let $A^{(1)} = \tilde{A}$ and use the B_1 associated with this $A^{(1)}$ to compute $r_i^{(1)}$ according to (3.2.2). We set $\tilde{r}_i^{(1)} = r_i^{(1)}$, then for $k = 1, 2, \ldots, n$ use (3.2.11), (3.2.10), (2.5.2), (2.2.3), and (2.5.4) to transform $A^{(k)}$ to $A^{(k+1)}$; and (3.2.16), (3.2.13), and (3.2.14) to update $\tilde{r}_i^{(k)}$'s to $\tilde{r}_i^{(k+1)}$'s. In this manner A has been transformed to an upper triangular matrix $A^{(n+1)} = \hat{U}$ (say), such that

$$\hat{U} = A^{(n+1)} = L_n P_n \cdots L_1 P_1 A^{(1)},$$

using (3.2.10) and (2.5.2),

$$= L_n P_n \cdots L_1 P_1 A Q,$$

using (3.2.9),

$$= \hat{L} A Q, \qquad \text{where} \quad \hat{L} = L_n P_n \cdots L_1 P_1.$$

Therefore,

$$A^{-1} = Q \hat{U}^{-1} \hat{L},$$

which is the factored form of A^{-1} given by (2.5.13).

In the next section we discuss how both P and Q in (3.2.1) can be determined a priori such that $\hat{A}$ is in a form in which either there is no fill-in or the fill-in is limited to only certain regions of $\hat{A}$.

3.3.   Desirable Forms for Gaussian Elimination

One of the most common forms which does not lead to any fill-in when the diagonal elements are chosen as pivots is the full band form, which is defined as follows.

> *DEFINITION*   A matrix A for which $a_{ij} = 0$, for $|i - j| > \beta$ is a *band matrix*. If in addition $a_{ij} \neq 0$ for all $|i - j| \leq \beta$ then it is a *full band matrix*. The quantity β is the *bandwidth* of A. Note that for a symmetric matrix having a locally variable bandwidth, which was defined in Section 1.3, $\beta = \max_i \theta_i$.

If A is a *full band matrix* and the pivots are chosen on the diagonal starting from the top left-hand corner, then in view of the proof of Theorem 2.5.5., there is no fill-in; for B_k is of full band form and there-fore, whenever $b_{ik}^{(k)} = b_{kj}^{(k)} = 1$, then $b_{ij}^{(k)} = 1$. On the other hand, if the band matrix A is not full, then some elements within the band are zero and the fill-in is limited to such elements within the band.

In order to discuss some of the desirable forms, we assume that A can be permuted to $\hat{A}$ by (3.2.1), where $\hat{A}$ is in the following form.

$$
(3.3.1) \qquad \hat{A} = \begin{bmatrix}
A_{11} & A_{12} & \cdots & A_{1,p-1} & A_{1p} \\
0 & A_{22} & \cdots & A_{2,p-1} & A_{2p} \\
0 & 0 & & & \\
\vdots & \vdots & & \vdots & \vdots \\
0 & 0 & & A_{p-1,p-1} & A_{p-1,p} \\
A_{p1} & A_{p2} & & A_{p,p-1} & A_{pp}
\end{bmatrix},
$$

where the submatrices on the diagonal, A_{ii}, $i = 1, 2, \ldots, p$, are non-singular matrices. If the pivots are chosen from the nonzero elements of the diagonal blocks A_{ii}, starting with A_{11} and proceeding sequentially, then the fill-in is limited to only those block matrices in (3.3.1) which are not labeled as zeros. In Section 3.10, we describe another order of choosing the pivots, such that no fill-in will take place even in the matrices A_{ji}, $j < i$, $i \neq p$.

In view of the above facts, we would like to determine two permutation matrices P and Q such that $\hat{A}$ in (3.2.1) has the form (3.3.1). If A is symmetric then it is generally advantageous to express $\hat{A}$ also in a symmetric form, since in this case only the nonzero elements of A on and above the leading diagonal need to be stored. Furthermore, in view of Theorem 2.5.19, if the diagonal elements can be chosen as pivots (e.g. when A is positive definite), then the symmetry is preserved during the elimination process and the lower triangular part is not stored. In this case instead of (3.2.1), we have

(3.3.2) $$PAP' = \hat{A},$$

and in (3.3.1), $A_{ij} = 0$, for all $i \neq j$, except $i = p$ or $j = p$.

In the following section we will describe several methods for the determination of P and Q in (3.2.1) and (3.3.2) that lead to the various desirable forms of $\hat{A}$ some of them given by (3.3.1). In order to discuss these methods, we will need some simple ideas from graph theory which are given in the next section. For additional details the reader is referred to Busacker and Saaty (1965), Harary (1969), Bellman *et al.* (1970).

3.4.   *Matrices and Graphs*

Let b_{ij} denote the (i, j) element of a matrix B which is obtained by replacing each nonzero element of A by unity. We associate with B, a *labeled graph* Ω, a *labeled digraph* (or directed graph) Ω_D, a *labeled bipartite graph* Ω_B, a *labeled row graph* Ω_R, and a *labeled column graph* Ω_C as follows.

DEFINITIONS

The *labeled graph* Ω consists of a set of n *vertices* labeled as $1, 2, \ldots, n$ and τ_0 *edges*. There is said to be an *edge* $[p, q]$ joining a vertex p to another vertex q if either b_{pq} or b_{qp} (or both) are equal to one. It follows that τ_0 is equal to the total number of nonzero elements of $B + B'$ that lie above the diagonal.

The *labeled digraph* (*directed graph*) Ω_D consists of a set of *n vertices* labeled as $1, 2, \ldots, n$ and τ_D *arcs*. There is said to be an *arc* $[p, q]$ from a vertex p to another vertex q iff $b_{pq} = 1$. Thus τ_D is equal to the total number of nondiagonal elements in B.

The *labeled bigraph* (*bipartite graph*) Ω_B consists of two distinct sets of vertices R and C, each set having n elements labeled $1, 2, \ldots, n$, together with τ *edges* which join the vertices in R to those in C. There is an *edge* $[p, q]$ with vertex p in set R and vertex q in set C iff $b_{pq} = 1$. Thus τ is the total number of nonzero elements in B.

The *labeled row graph* Ω_R and the *labeled column graph* Ω_C are, respectively, the *labeled graphs* of $B * B'$ and $B' * B$; where $*$ denotes that in computing the inner product of vectors in the above matrix multiplications only Boolean addition $\oplus$ has been used, namely, $1 \oplus 1 = 1$. Since $B * B'$ and $B' * B$ are both symmetric, therefore τ_R (the number of edges in Ω_R) and τ_C (the number of edges in Ω_C) are equal to the total number of ones that lie above the diagonals of $B * B'$ and $B' * B$, respectively.

In Fig. 3.4.1 the various graphs $\Omega, \Omega_D, \Omega_B, \Omega_R$, and Ω_C that correspond to a given matrix are shown.

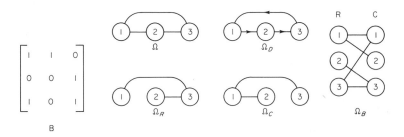

Fig. 3.4.1.   Labeled graphs associated with a matrix.

If the rows and columns of B are permuted such that all its diagonal elements remain on the diagonal, namely,

(3.4.1) $PBP' = \tilde{B}$

(P is a permutation matrix), then the only changes in the corresponding Ω and Ω_D are a relabeling of their vertices; otherwise they remain the same.

The application of different row and column permutations to B, that is,

$$(3.4.2) \qquad PBQ = \hat{B}$$

(P and Q are permutation matrices) leaves Ω_B unchanged except for a renumbering of the vertices in R and C.

Row permutations P on B lead to a renumbering of the vertices of Ω_R and column permutations Q to a renumbering of the vertices of Ω_C. From (3.4.2) we have

$$(3.4.3) \quad PB * B'P' = \hat{B} * \hat{B}' \qquad \text{and} \qquad Q'B' * BQ = \hat{B}' * \hat{B},$$

and therefore from the definitions of Ω_R and Ω_C it follows that Q and P in (3.4.2) have no effect on Ω_R and Ω_C, respectively.

If the vertices of $\Omega, \Omega_D, \Omega_B, \Omega_R$, and Ω_C are not labeled then the prefix *labeled* is dropped and they are, respectively, called *graph*, *digraph*, *bigraph*, *rowgraph*, and *column graph*. Thus B and $\tilde{B}$ in (3.4.1) are associated with the same graph, digraph, rowgraph and column graph. Matrices B and $\hat{B}$ in (3.4.2) have the same bigraph, rowgraph, and column graph. This invariance of graphs, digraphs, bigraphs, row-graphs, and column graphs under row and column permutations, makes them especially useful in studying the fill-in and in finding desirable forms for the Gaussian elimination. We shall need some additional definitions for this purpose. These apply to $\Omega, \Omega_D, \Omega_R$, and Ω_C but not to Ω_B. Those for Ω_B will be given later.

DEFINITIONS

The vertices p and q that are joined by an edge in Ω, Ω_R, or Ω_C, or an arc in Ω_D, are said to be *adjacent vertices*. If there exists a subset $v_1, v_2, \ldots, v_\sigma, v_{\sigma+1}$ of distinct vertices such that for $i = 1, 2, \ldots, \sigma$, the vertices v_i and v_{i+1} are adjacent, then v_1 and $v_{\sigma+1}$ are said to be *connected* by a *path* $[v_1, v_2, \ldots, v_{\sigma+1}]$ in Ω, Ω_R, or Ω_C (by a directed path in Ω_D) of *length* σ.

If in $[v_1, v_2, \ldots, v_{\sigma+1}]$ the *starting vertex* v_1 is the same as the *terminating vertex* $v_{\sigma+1}$, then the path is said to be a *cycle* of Ω, Ω_R, or Ω_C or a *directed cycle* of Ω_D of *length* σ.

Since in Ω, Ω_R, Ω_C, and Ω_D, the total number of vertices is n, the maximum value that σ can have is n.

If the set of vertices in $\Omega(\Omega_D, \Omega_R, \Omega_C)$ can be divided into two or more subsets, such that only the vertices within a subset are connected, then $\Omega(\Omega_D, \Omega_R, \Omega_C)$ is said to have two or more *disjoint labeled subgraphs*.

The number of edges of $\Omega(\Omega_R, \Omega_C, \Omega_B)$ in which a given vertex occurs is the *degree* of the vertex. Thus the degree of vertex i in R of Ω_B is the number of ones in the ith row of B. The degree of vertex i of Ω is the number of off-diagonal ones in the ith row of $B \oplus B'$. The number of arcs of Ω_D that start at a given vertex is the *out-degree* of the vertex. Thus the out-degree of vertex i is the total number of ones in row i of B.

The *in-degree* of the vertex is the number of arcs of Ω_D that terminate in the given vertex. Thus the in-degree of vertex j is the total number of off-diagonal ones in the jth column of B.

We shall now state and prove some theorems which we will need later on.

(3.4.4)   *THEOREM*   Let

(3.4.5) $$W = B \oplus B' \oplus I,$$

(3.4.6) $$W^{\sigma+1} = W^\sigma * W,$$

where $\sigma = 1, 2, \ldots, n - 1$, then $e_i' W^\sigma e_j = 1$, iff the vertices i and j of the labeled graph associated with B are connected by a path of length less than or equal to σ.

Proof By induction. The theorem is certainly true for $\sigma = 1$, since $e_i' W e_j = w_{ij} = 1$ iff there is an edge of length one between i and j. Suppose the theorem is true for a certain value of σ, then we will show that it is also true for $\sigma + 1$. From (3.4.6) we have

(3.4.7) $$w_{ij}^{\sigma+1} = \sum_{p=1}^n w_{ip}^\sigma * w_{pj}.$$

Now,

(3.4.8) $w_{ij}^{\sigma+1} = 1$ iff $w_{ip}^\sigma = 1$ and $w_{pj} = 1$ for at least one p.

But $w_{ip}^\sigma = 1$ iff vertices i and p are connected by a path of length σ or less and $w_{pj} = 1$ iff vertices p and j are connected

by an edge when $p \neq j$. When $p = j$ then $w_{jj} = 1$ in view of
(3.4.5). Therefore, in either case, (3.4.8) holds iff vertices i and j
are connected by a path of length less than or equal to $\sigma + 1$.
This completes the proof of the theorem.

Since the maximum length of a path is n (as there are n vertices), the
matrix W^n will be used in the following sections to determine all paths
and cycles. A theorem similar to Theorem 3.4.4 for labeled directed
graphs is

(3.4.9)   *THEOREM*   Let

(3.4.10) $$\hat{W} = B \oplus I,$$

(3.4.11) $$\hat{W}^{\sigma+1} = \hat{W}^{\sigma} * \hat{W}$$

where $\sigma = 1, 2, \ldots, n$ then $e_i' \hat{W}^{\sigma} e_j = 1$, if and only if in the
labeled digraph associated with B there is a directed path
from vertex i to vertex j of the labeled digraph associated with
B of length less than or equal to σ.

Proof   Same as that of Theorem (3.4.4), with W replaced by
$\hat{W}$.

3.5.   *The Block Diagonal Form*

In this section we shall describe some methods for transforming a
given matrix to *Block Diagonal Form* (BDF). We recall that $\hat{A}$ defined
by (3.3.1) is in BDF if for all $i \neq j$, $A_{ij} = 0$ and the diagonal blocks (A_{ii}s)
are square matrices. If $\hat{A}$ is in BDF then, in view of (3.2.1) and (3.4.2), $\hat{B}$
is also in *BDF*. This implies that both Ω_R and Ω_C associated with $\hat{B}$
consist of disjoint labeled subgraphs, and each labeled subgraph is
associated with a particular diagonal block. Since Ω_R (and Ω_C) associated with $\hat{B}$ and B are the same except for a relabeling of vertices, Ω_R and
Ω_C associated with B can be used to determine P and Q in (3.2.1) as
follows (Harary, 1962; Tewarson, 1967c).

(3.5.1)   *THEOREM*   If

(3.5.2) $$W = B * B'$$

and

(3.5.3) $$W^{2^{h+1}} = W^{2^h} * W^{2^h}, \qquad h = 0, 1, 2, \dots,$$

then there exists an h_1 such that

(3.5.4) $$W^{2^{h_1+1}} \equiv W^{2^{h_1}} = F$$

and $e_i' F e_j = 1$ iff row i and row j of B belong to the same diagonal block of $\hat{B}$.

Proof   If in Theorem 3.4.4, B is replaced by $B * B'$, then, in view of the fact that $B * B'$ is symmetric and has all its diagonal elements nonzero, equation (3.4.5) becomes

$$W = (B * B') \oplus (B * B')' \oplus I = B * B',$$

which is same as (3.5.2). Therefore, from the definition of row graph Ω_R and Theorem 3.4.4 it follows that if $e_i' W^{2^h} e_j = 1$, then there is a path between vertices i and j. If v is the size of the largest diagonal block of $\hat{B}$, then all paths in Ω_R are less than or equal to v and $v \leqslant n$. Therefore, for all h such that $2^h > v$, W^{2^h} remains unchanged and there exists an h_1 for which (3.5.4) is true and $e_i' W^{2^{h_1}} e_j = 1$ iff rows i and j of B belong to the same diagonal block in $\hat{B}$. This completes the proof of the theorem.

In order to make use of the above theorem, equation (3.5.3) is used until $2^{h_1} \geqslant n$, for generally we do not know v but only the fact that $v \leqslant n$. Let $F = W^{2^{h_1}}$. Instead of (3.5.3), there are several practical methods available for the determination of the matrix F (Baker, 1962; Warshall, 1962; Ingerman, 1962; Comstock, 1964). We shall briefly describe the one due to Comstock (1964):

A search in the first row of W is made until a zero is encountered, say in column j, then this zero is replaced by

$$\sum_{p=1}^{n} w_{1p} w_{pj} \qquad \text{(Boolean)}.$$

The search continues with columns $j + 1, j + 2, \ldots$ of row one until another zero is found, say in column q. This zero is replaced by

$$\sum_{p=1}^{n} w_{1p} w_{pq}.$$

This process is continued until a whole row has been searched. Then the remaining rows are also searched and modified in the same manner. When all the rows have been processed, one begins with the first row again. This processing is continued until a complete pass of the matrix has been made with no changes. The resulting matrix is F.

It is evident that $f_{ij} = e_i' F e_j = 1$ if and only if row i and j belong to the same diagonal block. Thus all the rows of B that correspond to non-zeros in the first column of F, belong to the first diagonal block. If all rows of F, for which $f_{i1} = 1$ are deleted then the next nonzero column of F can be used in the same manner as the first column of F, to find the rows of B which belong to the second diagonal block in $\hat{B}$, and so on. In this manner we can find the diagonal blocks of $\hat{B}$ to which each row of B belongs.

The following corollary to Theorem 3.5.1 can be used for assigning the columns of B which belong to different diagonal blocks.

(3.5.5)   COROLLARY   If F is as defined in Theorem 3.5.1 and

(3.5.6)
$$F * B = \bar{F}$$

and $\bar{f}_{ij} = e_i' \bar{F} e_j = 1$, then row i and column j of B belong to the same diagonal block of $\hat{B}$.

Proof   From (3.5.6) we have

$$\bar{f}_{ij} = \sum_{p=1}^{n} f_{ip} * b_{pj},$$

and $\bar{f}_{ij} = 1$, iff $f_{ip} = b_{pj} = 1$, for at least one value of p. From Theorem 3.5.1, we know that $f_{ip} = 1$ implies that rows i and p belong to the same diagonal block of $\hat{B}$, and since column j has at least one nonzero element in row p, therefore column j should also be in the same diagonal block as row p and row i. Thus $\bar{f}_{ij} = 1$ implies that row i and column j belong to the same diagonal block. This completes the proof of the corollary.

To make use of the above corollary we proceed as follows. For all columns of B that belong to the first diagonal block selected according to Theorem 3.5.1, we must have $\bar{f}_{ij} = 1$. Deleting or flagging such columns in $\bar{F}$, the next nonzero row of the resulting matrix gives the set of columns of B that belong to the second diagonal block and so on.

If in Theorem 3.5.1, instead of (3.5.2) we define $W = B' * B$, then we have a theorem for permuting the columns of B to $\hat{B}$. However, a slightly modified version, which uses Theorem 3.5.1 and thus avoids the duplication of work is as follows.

(3.5.7)   *THEOREM*   If F is as defined in Theorem 3.5.1, and

(3.5.8) $e_i'(B' * F * B)e_j = 1,$

then columns i and j of B belong to the same diagonal block of $\hat{B}$.

Proof   From (3.5.2), (3.5.3), and (3.5.4) we have

$$B' * F * B = B' * [(B * B') * (B * B') * \cdots * (B * B')] * B$$
$$= (B' * B) * (B' * B) * \cdots * (B' * B)$$
$$= (B' * B)^\sigma, \qquad \sigma > v.$$

Therefore from the definition of the column graph Ω_C, and the same arguments as in the proof of Theorem 3.5.1, it follows that

$$e_i'(B' * F * B)e_j = e_i'(B' * B)^\sigma e_j = 1$$

implies that columns i and j are connected by a path and therefore belong to the same diagonal block. This completes the proof of the theorem.

Since (3.5.8) requires one more matrix multiplication than (3.5.6), Corollary 3.5.5 is generally used instead of Theorem 3.5.7 for permuting the columns into block diagonal form. We give a simple example showing how Theorem 3.5.1 and Corollary 3.5.5 can be used to permute a given B to the block diagonal form $\hat{B}$. Let

$$B = \begin{bmatrix} 0 & 1 & 1 & 0 \\ 1 & 0 & 0 & 1 \\ 1 & 0 & 0 & 0 \\ 0 & 1 & 1 & 0 \end{bmatrix}, \quad \text{then} \quad W = \begin{bmatrix} 1 & 0 & 0 & 1 \\ 0 & 1 & 1 & 0 \\ 0 & 1 & 1 & 0 \\ 1 & 0 & 0 & 1 \end{bmatrix}$$

and

$$W^2 = \begin{bmatrix} 1 & 0 & 0 & 1 \\ 0 & 1 & 1 & 0 \\ 0 & 1 & 1 & 0 \\ 1 & 0 & 0 & 1 \end{bmatrix}.$$

Since $W^2 \equiv W$, therefore

$$F = W \quad \text{and} \quad F * B = W * B = \begin{bmatrix} 0 & 1 & 1 & 0 \\ 1 & 0 & 0 & 1 \\ 1 & 0 & 0 & 1 \\ 0 & 1 & 1 & 0 \end{bmatrix},$$

and from Theorem 3.5.1 it follows that the first and the last rows of B belong to the first diagonal block (since $f_{11} = w_{11} = f_{41} = w_{41} = 1$) and the rest of the rows to the second diagonal block. From Corollary 3.5.5 we conclude that the second and the third columns of B lie in the first diagonal block and rest of the columns in the second diagonal block. Therefore

$$P = \begin{bmatrix} 1 & 0 & 0 & 0 \\ 0 & 0 & 0 & 1 \\ 0 & 1 & 0 & 0 \\ 0 & 0 & 1 & 0 \end{bmatrix}, \quad Q = \begin{bmatrix} 0 & 0 & 1 & 0 \\ 1 & 0 & 0 & 0 \\ 0 & 1 & 0 & 0 \\ 0 & 0 & 0 & 1 \end{bmatrix},$$

and

$$\hat{B} = PBQ = \begin{bmatrix} \begin{array}{cc|cc} 1 & 1 & 0 & 0 \\ 1 & 1 & 0 & 0 \\ \hline 0 & 0 & 1 & 1 \\ 0 & 0 & 1 & 0 \end{array} \end{bmatrix}.$$

3.6.   The Block Triangular Form

If in $\hat{A}$ given by (3.3.1), $A_{pj} = 0$, $j \neq p$ and A_{ii}, $i = 1, 2, \cdots, p$ are square matrices, then $\hat{A}$ is said to be in a *Block Triangular Form* (BTF). In this section methods for permuting B (which is the same as permuting A) to BTF will be described. The first method consists of two parts such that

(3.6.1) $\hat{P}B\hat{Q} = \tilde{B}$

and

(3.6.2) $\tilde{P}\tilde{B}\tilde{P}' = \hat{B},$

where $\tilde{B}$ has all its diagonal elements equal to one and $\hat{B}$ is in BTF. From (3.6.1) and (3.6.2) we have

(3.6.3) $PBQ = \hat{B}$, $P = \tilde{P}\hat{P}$,   and $Q = \hat{Q}\tilde{P}'$.

Thus we want to find P and Q in (3.6.3).

If A is nonsingular there must exist at least one nonzero term in its determinant. This term consists of n entries of A of which no two appear either in the same row or column of A. Thus the rows and columns of A can be permuted independently to bring the nonzero elements in this term to the diagonal. The determination of $\hat{P}$ and $\hat{Q}$ in (3.6.1) such that $\tilde{b}_{ii} = 1$, for all $i = 1, 2, \ldots, n$ is done as follows (Steward, 1962, 1965; Dulmage and Mendelsohn, 1963; Kettler and Weil, 1969; Harary, 1971a; Duff, 1972).

(3.6.4)   *ALGORITHM*   Set

$$B = B_1, \quad V_1 = \sum_{i=1}^{n} e_i, \quad U_1 = V_1',$$

$$P_1 = Q_1 = I, \quad \text{and} \quad k = 1.$$

STEP 1

For all i, j with

(3.6.5) $b_{ij} = 1, \quad e_i'V_k = 1, \quad \text{and} \quad U_k e_j = 1,$

compute

(3.6.6) $e_{\alpha_k}' B V_k + U_k B e_{\beta_k} = \min_{i,j} (e_i' B V_k + U_k B e_j).$

If there is no i and j pair for which (3.6.5) holds then go to Step 2; otherwise, set both the α_k element of V_k and the β_k element of U_k to zero. Interchange the kth and the α_kth rows of P_k and the kth and the β_kth columns of Q_k.

Replace k by $k + 1$. If $k = n + 1$ go to Step 2, otherwise return to the beginning of this step.

> **Remarks**   For each value of k we select the (α_k, β_k) element of B to be the (k, k) element of $\tilde{B}$ in (3.6.1). Since all the elements that lie either in row α_k or column β_k of B cannot be chosen later as the diagonal elements of $\tilde{B}$, therefore (3.6.5) and (3.6.6) guarantee that a maximum number of nonzero elements of B will be available as possible choices for the rest of the diagonal elements of $\tilde{B}$. This in turn implies that at the termination of the current step we will generally have only a few rows and columns of B which have not been used in choosing the diagonal elements for $\tilde{B}$. If $k = n + 1$ then all the nonzero diagonal elements have been found.

STEP 2

Evaluate

(3.6.7) $B^{(k)} = P_k B Q_k.$

If $k = n + 1$ then set $\tilde{B} = B^{(k)}$ and go to Step 3; otherwise use the steps in the flow chart in Fig. 3.6.1 on matrix $B^{(k)}$ to permute it to a matrix $B^{(n+1)}$ which has all ones on the diagonal.

> ***Remarks on Fig. 3.6.1***   Note that if we arrive at box numbered 12, then the unflagged rows do not contain any ones in the flagged columns. Since the number of flagged columns

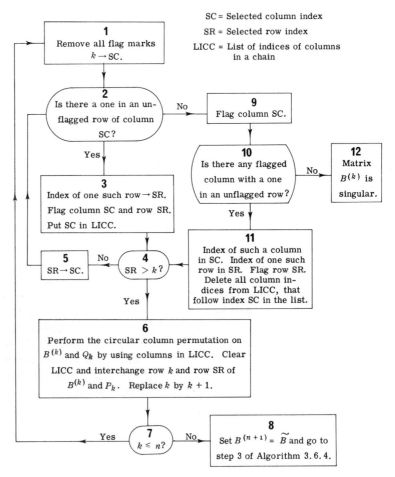

Fig. 3.6.1.   Flow chart for creating a diagonal of all ones in $B^{(k)}$.

is one more than flagged rows at this time, the matrix is singular. Thus for nonsingular matrices we cannot get to box 12, and it is always possible to get a matrix $B^{(n+1)}$ with all ones on the diagonal. The matrix $B^{(k)}$ has all zeros in the southeast corner, that is, $e_i'B^{(k)}e_j = 0$ for all $i, j \geqslant k$. The transformation of $B^{(k)}$ to $B^{(k+1)}$ involves first finding a chain from a one in the kth column in the northeast corner to a one in some row in the southwest corner of $B^{(k)}$ as shown in Fig. 3.6.2. This list of columns is kept in LICC.

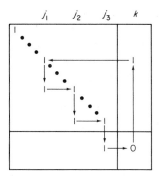

Fig. 3.6.2.   Chain from northeast to southwest corner.

Then a circular column shift $k \to j_1 \to j_2 \to j_3 \to k$ is performed to bring the one in the southwest corner to the southeast corner. If at any time a one cannot be found in an unflagged row of a column under consideration, then we search for another set of columns to get a chain by using boxes 10 and 11 in Fig. 3.6.1 until a column with a one in the southwest corner is found. The flagging of the rows and the columns of $B^{(k)}$ assures that a path that leads to a column with no element in its unflagged rows is not repeated again and the chain is determined in a finite number of steps. Note that LICC contains a list of indices of columns of $B^{(k)}$ which form a chain from a one in the kth column to a one in the southwest corner and therefore has to be modified in box 11 to get rid of those columns that lead to a dead end in the search for a chain.

STEP 3

Set $P_k = \hat{P}$ and $Q_k = \hat{Q}$. $\tilde{B}$ has all ones on the diagonal and the algorithm is complete.

> *Remark* The determination of $\hat{P}$ and $\hat{Q}$ such that $\tilde{B}$ has all nonzero diagonal elements, is equivalent to a renumbering of the vertices of the labeled bigraph Ω_B corresponding to B, such that any two vertices in R and C having the same number as labels are adjacent (connected by an edge). A string of one bits of maximal length extending down the principal diagonal of a matrix is called the *maximal transversal* (Dulmage and Mendelsohn, 1963). We recall that for a nonsingular matrix of order n, the length of this maximal transversal is n.

The determination of $\tilde{P}$ in (3.6.2), which constitutes the second part of the first method for transforming B to the BTF $\hat{B}$, is based on the following theorem.

(3.6.8)   *THEOREM*   If all the diagonal elements of B are ones and

(3.6.9) $$B^{2^{h+1}} = B^{2^h} * B^{2^h}$$

then there exists an h_1 such that

(3.6.10) $$B^{2^{h_1+1}} = B^{2^{h_1}} = F,$$

> and $e_i' F e_j = 1$ iff there is a directed path from vertex i to vertex j in Ω_D (the labeled digraph associated with B).

> *Proof* Since $B \oplus I = B$, it follows from Theorem 3.4.9 that $e_i' B^v e_j = 1$ iff there is a directed path of length v or less between vertices i and j of Ω_D. If v is the maximum length of the directed path between any two vertices of Ω_D, then $v \leqslant n$, and evidently for all $2^h \geqslant v$, the matrix B^{2^h} remains unchanged. Therefore, there exists an h_1 satisfying (3.6.10) and it follows that $e_i' F e_j = 1$ iff there is a directed path from vertex i to vertex j. This completes the proof of the theorem.

If in Theorem 3.6.8 instead of B we start with the $\widetilde{B}$ that was obtained from B in Algorithm 3.6.4, and then compute F either according to (3.6.9) and (3.6.10) or by the Comstock's method described after the proof of Theorem 3.5.1, then

$$e_i'Fe_j = e_j'Fe_i = 1$$

implies that the vertices i and j lie in the same directed cycle. Since the vertices of Ω_D that lie in the same directed cycle belong to the same diagonal block in $\widehat{B}$, therefore we can use F to determine $\widehat{B}$ from $\widetilde{B}$ as follows (Harary, 1962):

Determine all j such that

$$e_1'Fe_j = e_j'Fe_1 = 1,$$

then all rows and columns of $\widetilde{B}$ with such indices j belong to the same diagonal block of $\widehat{B}$. Delete (or flag) all such rows and columns of F. The next undeleted (or unflagged) row and the corresponding column can now be used in exactly the same manner as the first row and the first column were used to determine the rows and columns for another diagonal block and so on. In this manner all the rows and columns of $\widetilde{B}$ can be associated with the relevant diagonal blocks. Note that if $F = F'$ then $\widetilde{B}$ can be permuted to a BDF. (This would be an alternative, though somewhat slower, way than Section 3.5.) In order to arrange the diagonal blocks such that $\widehat{B}$ is in BTF, we proceed as follows: With each diagonal block we associate a row and a column which are formed respectively by taking the Boolean sum of all rows and columns of F that are associated with the diagonal block. In this way we get from F a new square matrix $\widehat{F}$ having the same order as the number of diagonal blocks, the labeled digraph corresponding to F is called the *labeled condensation digraph* Ω_S of $\widetilde{B}$. Its vertices are the diagonal blocks. Since a condensation graph remains unchanged by a relabeling of its vertices, $\widehat{B}$ has the same condensation graph. A vertex from which only arcs emanate is called an *emitter* and a vertex not connected to any other vertices is an *isolated* vertex. Thus Ω_S must have an emitter ($\widehat{B}$ has no nonzero blocks below the diagonal) or an isolated vertex. We call the diagonal block corresponding to this vertex of Ω_S as the first diagonal block. Dropping the row and the column of $\widehat{F}$ that correspond to the choice for the first diagonal block from consideration and deleting the corresponding vertex and arcs of Ω_S, we find that the resulting Ω_S (and F) must also have an emitter or an isolated vertex and the corresponding diagonal block is taken as the second diagonal block and so on. Thus the order of diagonal blocks of $\tilde{B}$ is determined. A record of all these permutations is kept and the final result is denoted by $\tilde{P}$, then using (3.6.2) the block upper triangular matrix $\hat{B}$ is determined.

The second method for permuting B to BTF is identical to the first method as far as the determination of a maximum transversal is concerned, but its second part is different. The difference is based on the observations below.

It is possible to avoid the computation of the matrix F given by (3.6.10) when transforming $\tilde{B}$ to the block triangular form $\hat{B}$ as follows (Steward, 1965).

First, we note that the diagonal blocks of $\hat{B}$, which are of order two or more, correspond to the cycles of the directed graph associated with it. On the other hand, all the columns of $\hat{B}$ having no nonzero element below the diagonal and that lie before its first diagonal block can easily be found from $\tilde{B}$ in the following way.

STEP 1

We determine a column of $\tilde{B}$ having only a single one and move this column and the corresponding row (which may have several ones) to the northwest corner and flag this column and the corresponding row. This process is repeated until there are no more columns with single ones in the unflagged rows among the unflagged columns of $\tilde{B}$.

STEP 2

We determine the columns (and rows) which lie in the same diagonal block as the first unflagged column of $\tilde{B}$: From Step 1 it follows that the first unflagged column of $\tilde{B}$ has at least one off-diagonal one in an unflagged row. The column corresponding to the row of the first such one must also have a one in an unflagged row, and so on. At some time, a previously mentioned column is encountered, thus completing a directed cycle. We replace the set of columns in a directed cycle by one column which is the Boolean sum of all such columns without their diagonal elements.

This is called *collapsing* the columns in a directed cycle. The *Boolean sum of vectors* is the usual vector addition with $1 \oplus 1 = 1$. Similarly, the rows in the directed cycle are replaced by one row which is their Boolean sum. When a column with only a single one is flagged, then the columns which were collapsed to form it belong to the same block. The order in which the columns with single ones were flagged gives an order in which the diagonal blocks lie. We continue with Steps 1 and 2 until all the columns and rows of the matrix have been flagged.

	1	2	3	4	5	6	7
1	X	X	·	·	·	X	·
2	X	X	·	·	·	·	·
3	·	·	X	·	X	·	X
4	·	·	·	X	·	·	·
5	·	·	·	X	X	·	·
6	·	X	·	·	X	X	·
7	X	·	·	·	·	·	X

Fig. 3.6.3

We clarify these rules by beginning with Fig. 3.6.3 and proceeding step by step until we finish with Fig. 3.6.4. The positions of ones are indicated by ×, zeros by ·, and any ones created by collapsing rows or columns by +.

1. Column 3 has only a single one. Flag column 3 and row 3. Put a 1 in row 3 of the "order" column to record the sequence in which the column was eliminated.

2. Column 7 has only a single one in its unflagged rows. Flag column 7 and row 7 and put a 2 in the "order" column.

Order	into		1	2	3	4	5	6	7
3		1	X	X	•	•	•	X	•
	1	2	X	X	•	•	•	•	•
	1	3	•	•	X	•	X	•	X
5		4	•	•	•	X	•	•	•
4		5	•	•	•	X	X	•	•
	1	6	+	X	•	•	X	X	•
2		7	X	•	•	•	•	•	X

Fig. 3.6.4

3. There are no more unflagged columns with single ones in the unflagged rows. We begin by tracing a directed path with column 1, choosing always the first one encountered in the column. This gives the directed path $[1, 2, 1]$ which implies that 1 and 2 are in a directed cycle. We collapse column 2 into column 1 by adding $+$'s to column 1 to make it the Boolean sum of column 1 with the nondiagonal ones of column 2. Similarly, collapse row 2 into row 1, and flag column 2 and row 2. Put 1 in the second row of the "into" column to indicate that column 2 has been collapsed into column 1.

4. Beginning with column 1 we trace the directed path $[1, 6, 1]$ and collapse column 6 into column 1 and row 6 into row 1. Flag column 6 and row 6 and put a 1 in row 6 of "into" column.

5. Column 1 has now only a single one in unflagged rows. Therefore, we put a 3 in the first row of the "order" column and flag row 1 and column 1.

6. Column 5 has a single one in its unflagged rows and we place a 4 in the "order" column and flag row 5 and column 5.

7. Column 4 has only a single one in unflagged rows, therefore, we put a 5 in the "order" column and flag column 4 and row 4.

There are now no more unflagged columns and rows and we can read off from the "order" and "into" columns the arrangement of diagonal blocks. The first block contains the $(3, 3)$ element of $\tilde{B}$, the second the $(7, 7)$ element. The third block contains the $(1, 1)$, $(2, 2)$, and $(6, 6)$ elements of $\tilde{B}$ as diagonal elements. The fourth and fifth blocks consists of $(5, 5)$ and $(4, 4)$ elements, respectively. Thus the permutation matrix $\tilde{P}$ of equation (3.6.2) is given by

$$\tilde{P} = (e_3, e_7, e_1, e_2, e_6, e_5, e_4)$$

and $\hat{B}$ by Fig. 3.6.5.

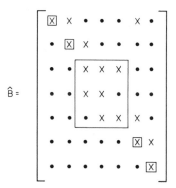

Fig. 3.6.5

If the matrix $\tilde{B}$ is small in size then its digraph can be used directly to obtain $\hat{B}$. The digraph corresponding to the matrix $\tilde{B}$ of Fig. 3.6.3 is given in Fig. 3.6.6. We look for a vertex that is an emitter (at which

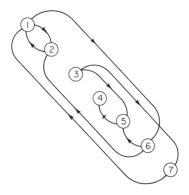

Fig. 3.6.6

no arcs terminate); vertex 3 is an emitter, we renumber it as 1 and then delete all arcs emanating from it. Now 7 becomes an emitter, renumber it 2 and delete the arc starting from it. There are no other emitters now. Vertices 1, 2, and 6 lie in a cycle, and we renumber them 3, 4, and 5, respectively, and delete all arcs which start from any one of them. Vertex 5 is now an emitter, we call it 6 and finally renumber 4 as 7. Thus

$$\tilde{P} = (e_3, e_7, e_1, e_2, e_6, e_5, e_4)$$

as before.

There are two other methods for transforming a matrix to a BTF or a form that is similar to it. These are discussed in Section 3.7.

3.7. *The Band Triangular Form*

A matrix B is said to be in a *Band Triangular Form* (BNTF) if there exists a β such that $0 < \beta \ll n$ and $b_{ij} = 0$, for all $i - j > \beta$, and this value of β cannot be decreased by row and column permutations on B.

Let the first nonzero element in the ith row of B be in the q_ith column, then we define

$$(3.7.1) \qquad\qquad \beta_i = i - q_i, \qquad q_i \leqslant i,$$
$$\qquad\qquad\qquad = 0, \qquad\quad q_i > i.$$

It is evident that $\max_i \beta_i = \beta$, if B is in BNTF. Note that the block triangular form (BTF) described in the last section is a special case of BNTF ($\beta + 1$ is the size of the largest diagonal block of B in this case).

In this section we assume that, if possible, the given matrix has already been transformed to BDF according to the methods in Section 3.5. If this was done then we consider the diagonal blocks of B one at a time. Therefore, there is no loss of generality, if we assume that the graph of B is connected.

FIRST METHOD FOR PERMUTING B TO BNTF

Let P and Q be permutation matrices such that

$$(3.7.2) \qquad\qquad\qquad \hat{B} = PBQ,$$

where the ith row and the jth column of B become, respectively, the ρ_ith row and the μ_jth column of $\hat{B}$. Our problem is to find those permutation matrices P and Q which minimize

$$\max_i \phi_i,$$

where ϕ_i is the distance from the leading diagonal of the leftmost nonzero element in row ρ_i of $\hat{B}$. Note that $\phi_i = 0$, if there is no nonzero element to the left of the diagonal in row ρ_i of $\hat{B}$.

This problem can be solved as follows (Chen, 1972). Let the nonzero elements of row i of B lie in columns $j_{i\alpha}$, $\alpha = 1, 2, \ldots, r_i$; where r_i is the total number of nonzero elements in row i of B. If the leftmost element in row ρ_i of $\hat{B}$ lies in column $\mu_{j_{is}}$, then

$$\rho_i - \mu_{j_{is}} = \phi_i, \qquad \rho_i \geqslant \mu_{j_{is}},$$

and for all $\mu_{j_{i\alpha}}$ not on the right of column ρ_i, $\rho_i - \mu_{j_{i\alpha}} \leqslant \phi_i$. If we prescribe that $\phi_i \geqslant 0$, for all i, then for any nonzero element to the right of the diagonal, say in column $\mu_{j_{it}}$

$$\rho_i - \mu_{j_{it}} < 0, \qquad \text{since} \quad \rho_i < \mu_{j_{it}};$$

therefore we have

$$\rho_i - \mu_{j_{i\alpha}} \leqslant \phi_i \qquad \text{for all} \quad \alpha.$$

In view of the above facts we have the following problem:

Corresponding to all $b_{ij} = 1$, find the set of integers ρ_i, μ_j, and ϕ_i which

(3.7.3) $$\text{minimize} \quad \max_i \phi_i$$

subject to the constraints

(3.7.4) $$1 \leqslant \rho_i, \quad \mu_j \leqslant n; \qquad 0 \leqslant \phi_i \leqslant n - 1$$

(3.7.5) $$\rho_i - \mu_{j_{i\alpha}} - \phi_i \leqslant 0; \qquad \alpha = 1, 2, \ldots, r_i$$

and

(3.7.6) $$\rho_i \neq \rho_j, \quad \mu_i \neq \mu_j \qquad \text{for} \quad i \neq j.$$

The above constrained Chebyshev problem can be expressed as an integer programming problem (Rabinowitz, 1968) and then solved in the usual manner (Dantzig, 1963a).

SECOND METHOD FOR PERMUTING B TO BNTF

This method is based on the use of row and column measures that are obtained on the basis of probabilistic considerations. This method does not minimize β, but yields a value which is reasonably close to the minimum. It is, however, much simpler than the first method given by equations (3.7.3) to (3.7.6).

The row and column measures which are used in this method to transform B to BNTF are derived as follows: Let $r_i = r_i^{(1)}$, $c_j = c_j^{(1)}$, where $r_i^{(1)}$ and $c_j^{(1)}$ are defined by (3.2.2) with $B_1 = B$. For a row i of B, let Λ_i be the set of all columns for which $b_{ij} = 1$, then we define

$(3.7.7)$ $$d_i = \sum_{j \in \Lambda_i} c_j.$$

For the purposes of the following discussion let us assume that B is already in a BNTF and the ones in the region $j \geqslant i - \beta$ are randomly distributed, namely, each element in the region has the same probability of being nonzero. Let $E(\theta)$ denote the expected value of θ, then the $E(r_i)$s will in general decrease as i increases from 1 to n. On the other hand, the $E(c_j)$s will generally increase with increasing js. In view of (3.7.7), for a given i, the average value of the c_js with $j \in \Lambda_i$ is d_i/r_i. It is easy to see that the $E(d_i/r_i)$s generally increase as i increases. Now for a given row i, the standard deviation π_i of the c_js, $j \in \Lambda_i$ is given by

$$\pi_i^2 = \frac{\sum_j c_j^2}{r_i} - \left(\frac{\sum_j c_j}{r_i}\right)^2, \qquad j \in \Lambda_i$$

which in view of (3.7.7) gives

$(3.7.8)$ $$\pi_i^2 = \frac{\sum_j c_j^2}{r_i} - \left(\frac{d_i}{r_i}\right)^2, \qquad j \in \Lambda_i.$$

In view of the characteristics of B, which we have assumed in the present discussion, it is not difficult to see that the expected values of π_is will generally decrease as i increases. We have also seen earlier in the discussion that as i increases the $E(r_i)$s and $E(r_i/d_i)$s generally decrease. Therefore, a reasonable measure M_i that we can associate with row i by taking all the three quantities r_i, π_i, and r_i/d_i into consideration is given by

(3.7.9) $$M_i = r_i + \delta\left(\frac{r_i}{d_i}\right) + \psi\pi_i$$

where δ and ψ are some numbers chosen in such a manner that r_i, $\delta(r_i/d_i)$, and $\psi\pi_i$ are of equal order of magnitude. Evidently the $E(M_i)$ s will generally decrease as i increases.

The column measures M_j s can also be determined in the same manner as the M_i s in (3.7.9), namely,

(3.7.10) $$M_j = c_j + \delta\left(\frac{c_j}{d_j}\right) + \psi\pi_j$$

where the d_j is defined in a manner similar to d_i in (3.7.7) as follows

(3.7.11) $$d_j = \sum_i r_i \quad \text{for all} \quad i \quad \text{with} \quad b_{ij} = 1,$$

and π_j is given by the following equation which is similar to (3.7.8),

(3.7.12) $$\pi_j^2 = \frac{\sum_i r_i^2}{c_j} - \left(\frac{d_j}{c_j}\right)^2 \quad \text{for all} \quad i \quad \text{with} \quad b_{ij} = 1.$$

The $E(M_j)$s will generally increase as j increases.

We have seen above that if B was in sufficiently BNTF then as i and j increase, the $E(M_i)$s decrease while the $E(M_j)$s increase. Now suppose that B is any given sparse matrix, not necessarily in BNTF, the non-zero elements of which are randomly distributed. If we compute the M_i s and M_j s for such a matrix B and then arrange its rows in descending values of M_i s and the columns in ascending values of M_j s, then we can expect that the permuted matrix $\hat{B}$ will be reasonably close to a BNTF.

We still have to determine the values of δ and ψ in (3.7.9) and (3.7.10) in case of a given sparse matrix B (the nonzero elements of B are randomly distributed throughout the whole matrix). We have found that in practice, taking $\delta = \tau^2/n^2$ and $\psi = 2$ (τ is the total number of nonzero elements in B), seems to work quite well (Tewarson, 1967c). A heuristic justification of the above choice for δ and ψ can be given if we tacitly assume that, for the complete matrix B,

$$E(r_i) \approx E(c_j) \approx \tau/n,$$

$$E(d_i) \approx E(d_j) \approx E(r_i)E(c_j) \approx \tau^2/n^2,$$

$$E\left(\frac{r_i}{d_i}\right) = E(r_i)/E(d_i),$$

and

$$2\pi_i = E(c_j) - \min c_j, \qquad \text{for all} \quad j \quad \text{with} \quad b_{ij} = 1.$$

Thus

$$2\pi_i \approx \frac{\tau}{n} \qquad \text{and} \qquad \frac{\tau^2}{n^2}E\left(\frac{r_i}{d_i}\right) \approx \frac{\tau}{n}$$

and therefore

$$\psi = 2 \qquad \text{and} \qquad \delta = \frac{\tau^2}{n^2}.$$

Once B has been permuted to BNTF, then in certain cases it is possible to perform additional row permutations to create *triangular corners* below the diagonal.

The triangular corners are defined as follows: If in matrix B, $q_1 \leqslant p_1 < p_2$ and $q_1 < q_2 \leqslant p_2$ and $b_{ij} = 0$ for $i \geqslant p_1$ and $j < q_1$, or $i > p_2$ and $q_1 \leqslant j \leqslant q_2$, then the triangular region with vertices (p_1, q_1), (p_2, q_1), and (p_2, q_2) is called a *triangular corner* (see Fig. 3.7.1).

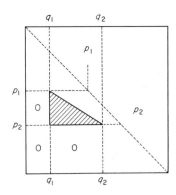

Fig. 3.7.1.   Triangular corner.

If $p_1 = q_1$ and $p_2 = q_2$, then the triangular corner is half of a diagonal block.

We recall the definition of β_i given by (3.7.1) and also the fact that $\max_i \beta_i \geqslant \beta$ for all matrices obtained from B by row–column permutations. If B is in BNTF then let Λ_m denote the nonempty set of row indices i, for which $\beta_i = \beta$. We now have the following theorem.

(3.7.13)   *THEOREM*   If B is in BNTF and $\beta_{i_2} - \beta_{i_1} > i_2 - i_1$ for some $i_2 > i_1$, then either i_2 does not belong to the set Λ_m or Λ_m consists of at least one additional row index.

     Proof   Let us interchange the rows i_1 and i_2 of B and denote the new values of β_{i_1} and β_{i_2} by $\hat{\beta}_{i_1}$ and $\hat{\beta}_{i_2}$, respectively. Then

$$\hat{\beta}_{i_1} = \beta_{i_2} - (i_2 - i_1)$$

which, in view of the fact that $i_2 > i_1$, implies

(3.7.14) $$\hat{\beta}_{i_1} < \beta_{i_2}.$$

     Also

$$\hat{\beta}_{i_2} = \beta_{i_1} + (i_2 - i_1),$$

which, in view of the inequality $\beta_{i_2} - \beta_{i_1} > i_2 - i_1$, gives

(3.7.15) $$\hat{\beta}_{i_2} < \beta_{i_2}.$$

If i_2 was the only element in Λ_m, then in view of (3.7.14) and (3.7.15), it is clear that we have decreased the value of β which is impossible since B was assumed to be in BNTF. Therefore either i_2 does not belong to Λ_m or if it does then there is at least one other row index in Λ_m to keep β same. This completes the proof of the theorem.

The above theorem can be used to permute the rows of B until for all

$$i_2 > i_1, \qquad \beta_{i_2} - \beta_{i_1} \leqslant i_2 - i_1.$$

At this stage, there may be several triangular corners, because whenever $i_1 < i_2$ and

(3.7.16) $$\beta_{i_2} - \beta_{i_1} = i_2 - i_1,$$

then for all $i_1 \leqslant i \leqslant i_2, \beta_{i_2} - \beta_i = i_2 - i$. If p_2 and p_1 are the maximum and minimum values of i_2 and i_1, respectively for which (3.7.16) holds and the leftmost ones in rows p_2 and $p_2 + 1$ are in column q_1 and $q_2 + 1$, respectively, then the triangle with vertices (p_1, q_1), (p_2, q_1) and (p_2, q_2) is a triangular corner (see Fig. 3.7.1). Therefore, we can use the methods of this section to permute B first to a BNTF and then use row interchanges to get as many triangular corners below the diagonal as possible. If there are one or more triangular corners with $p_1 = q_1$ and $p_2 = q_2$, then $\hat{B}$ is in a BTF as shown in Fig. 3.7.2. All elements in the

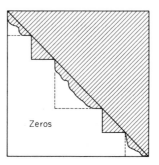

Fig. 3.7.2.   Block triangular form.

nonhorizontal parts of the boundary of the shaded area that lies below the diagonal are ones. All elements to the left of the boundary are zeros but to its right and in its horizontal parts itself there are both zeros and ones. The dotted lines are used to indicate the diagonal blocks.

We conclude this section with the remark that the above-mentioned row interchanges can be used even in those cases when B is only in an approximate BNTF. In this case $\max_i \beta_i$ may decrease in view of Theorem 3.7.13, and we not only create triangular corners but in certain cases get closer to the BNTF.

3.8.   *The Band Form*

In many applications A is symmetric and positive definite and therefore it is generally advantageous to choose the pivots on the diagonal to preserve symmetry, because, in view of Theorem 2.5.19, only the nonzero elements of the matrix above the diagonal need to be stored. In this case the same row and column permutations are performed on A to permute it to a desirable form for Gaussian elimination. This can be thought of as a permutation of the diagonal elements of A. The relevant equations are (3.3.2) and

$(3.8.1)$ $$PBP' = \hat{B}.$$

If we let

$(3.8.2)$ $$\beta_i = i - q_i, \qquad q_i \leqslant i,$$

where $\hat{b}_{iq_i}$ is the leftmost one in row i of $\hat{B}$, then we want to determine the minimum bandwidth β and a value of P, such that

$(3.8.3)$ $$\beta = \min_P \max_i \beta_i.$$

This problem can be expressed as a Chebyshev problem in a manner which is similar to the first method for permuting B to BNTF given in the preceding section, we shall not describe it here as it is generally not suitable for practical purposes. However, we shall describe below four practical methods for permuting the rows and columns of a given matrix to transform it to a *Band Form* (BF). For matrices that can be permuted to a matrix having no zeros within the matrix band, these methods yield such a matrix, in other cases they lead to a matrix with its $\max_i \beta_i$, reasonably close to the minimum β.

FIRST METHOD

This method is especially useful in problems where the original numbering of the a spatial system of vertices of a labeled graph determines the band width of the corresponding matrix (for example, finite element structural systems, lumped model thermal networks, electrical networks, pipe flow conduit systems, finite differences grid network systems). A simple example illustrating the decrease in bandwidth of a matrix by a renumbering of the vertices of the associated graph is given in Fig. 3.8.1. The vertices of the graph Ω which correspond to a matrix B are renumbered to yield graph $\hat{\Omega}$ which has $\hat{B}$ as the associated matrix.

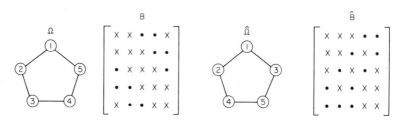

Fig. 3.8.1. Renumbering the vertices to decrease the bandwidth.

In B and $\hat{B}$, the zeros and ones are denoted by dots and $\times$'s, respectively. Note that the bandwidth of B is 4 but that of $\hat{B}$ is only 2. The method for renumbering the vertices of a graph in order to reduce the bandwidth of the corresponding matrix can be described in an algorithmic form as follows (Rosen, 1968).

1. Put integers $1, 2, \ldots, n$ in a *Vertex List* (VL) with n cells. Determine the maximum bandwidth of B and the two vertices that lead to it. If there are ties choose the first pair of vertices. (If $\max_i \beta_i = \beta_p$, then p and $p - \beta_p$ constitute the vertex pair). If the higher numbered vertex can be interchanged with a lowered numbered vertex to reduce the bandwidth ($\beta_i + (p - i) < \beta_p$ for some $i < p$), then go to Step 7.

2. If the lower numbered vertex can be interchanged with a higher numbered vertex to reduce the bandwidth then go to Step 7.

3. If a higher numbered vertex can be interchanged with a lower numbered vertex and still maintain the same bandwidth go to Step 6.

4. If a lower numbered vertex can be interchanged with a higher numbered vertex and still maintain the same bandwidth go to Step 6. (Remark: The purpose of Steps 3 and 4 is to rearrange the vertex pattern without increasing the bandwidth so that additional successful interchanges may be performed in Steps 1 and 2.)

5. No further interchanges are possible, and the matrix B is in band form now and for $j = 1, 2, \ldots, n, e_{\mathrm{VL}(j)}$ is the jth column of P, where VL(j) is the integer in the jth cell of VL. Then $\hat{A} = PAP'$. Stop.

6. If a maximum number of continuous interchanges have been performed in Steps 3 and 4 or two vertices which have already been interchanged are selected again for interchange then go to Step 5.

7. Perform the indicated vertex interchange, namely, interchange the corresponding elements of VL and the rows and columns of B and go to Step 1.

SECOND METHOD

Another vertex renumbering scheme which reduces the bandwidth of the corresponding matrix can be described in the following algorithmic form (Cuthill and McKee, 1969).

1. For each vertex i of Ω associated with the given matrix B, compute its degree ρ_i, where ρ_i is equal to the total number of off-diagonal ones in the ith row of B. Then choose any one vertex i_1 such that $\rho_{i_1} = \min_i \rho_i$ and label this vertex as one. In Fig. 3.8.2 the vertex 1* is relabeled as 1.

2. Renumber the vertices adjacent to vertex one in sequence beginning with 2 in the order of their increasing degree. (If there

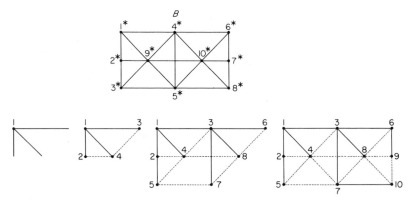

Fig. 3.8.2. Example of a vertex renumbering scheme.

are ties choose the first one). These vertices are said to be at first level. In Fig. 3.8.2 vertices 2*, 4*, and 9* are labeled 2, 3, and 4, respectively.

3. Repeat this procedure for each vertex at first level in sequence, that is, first for vertex 2, then for 3, and so on. In Fig. 3.8.2 the vertex adjacent to vertex 2 which has not already been numbered is 3*, it becomes vertex 5; adjacent to 3 (which was originally 4*) are vertices 6*, 5*, and 10*. Vertex 6* has a lower degree than vertex 5* which is of lower degree than vertex 10*; therefore number them as 6, 7, and 8, respectively. Note that vertices 5, 6, 7, and 8 have a path of length two from vertex 1 and they are said to be at second level.

4. Repeat the above procedure for vertices at each successive level until all the n vertices of Ω have been renumbered. (If Ω consists of two or more disjoint subgraphs then the procedure terminates as soon as all vertices in a subgraph have been renumbered. In this case, choose a starting vertex in each of the disjoint subgraphs and repeat Steps 2, 3, 4 for each of them.)

5. Finally, permute the rows and columns of B (or A) according to the renumbering of the vertices to get $\hat{B}$ (or $\hat{A}$). Matrices B and $\hat{B}$ are given in Fig. 3.8.3, they correspond to the graph and its renumbered form of Fig. 3.8.2. Note that bandwidths of B and $\hat{B}$ are, respectively, 8 and 5.

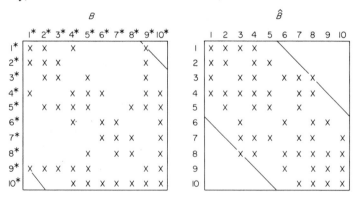

Fig. 3.8.3.   Matrix B and its permuted form $\hat{B}$.

The renumbering scheme described above does not, in general, lead to a minimum bandwidth. But it is possible to generate other sets for renumbering the vertices by starting with a different vertex i_1 in Step 1 such that

$$\rho_{\min} \leqslant \rho_{i_1} \leqslant \rho_{\min} + \tfrac{1}{2}\rho_{\max},$$

where $\rho_{\min}$ and $\rho_{\max}$ are, respectively, the minimum and the maximum values for ρ_i, and in case of ties at any of the succeeding steps, choosing a different order of numbering the vertices at that level. The set which leads to least bandwidth is then chosen to permute B to $\hat{B}$. Of course, it is possible to generate only a few of these sets, otherwise the whole process takes an unreasonable amount of time.

THIRD METHOD

. An iterative program that minimizes the average bandwidth

$$\bar{\beta} = \frac{1}{n} \sum_{i=1}^{n} \beta_i,$$

where β_i is defined as in (3.8.2) and is based on row (column) interchanges as in the first method, can be described in the following algorithmic form (Akyuz and Utku, 1968).

1. Interchange two successive rows (and the corresponding columns) of B if (a) $\bar{\beta}$ decreases, (b) $\bar{\beta}$ remains the same but the row having more zeros within the band goes away from the central row of B. Keep a record of interchanges in *Row Interchange* list RI (RI has initially the integers $1, 2, \ldots, n$, and every time a row interchange is performed the corresponding elements of RI are also interchanged.)

2. Perform the interchanges by comparing rows in complete sweeps. (A complete sweep consists of $n - 1$ steps. In each step two successive rows are compared and an interchange is made if the criteria in Step 1 are satisfied. The steps are performed in the following order: $(1, 2)$, $(n, n - 1)$, $(2, 3)$, $(n - 1, n - 2), \ldots$, up to the central row).

3. Stop, when no interchange is performed or no decrease in $\bar{\beta}$ occurs in an empirically determined number $(3 + n/100)$ of sweeps. For $j = 1, 2, \ldots, n, e_{\mathrm{RI}(j)}$ is the jth column of P, where $\mathrm{RI}(j)$ is the integer in the jth cell of the row interchange list RI.

FOURTH METHOD FOR REDUCING THE BANDWIDTH

Let the usual inner product (not Boolean) of two rows of a given matrix B called the *length of their intersection*. Let v_i denote the sum of the lengths of the intersections of row i with all the rows of B. If we assume that B is in a band form with bandwidth β then the number of ones in the shaded areas in Fig. 3.8.4 represent the values of v_i for three

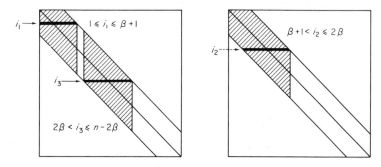

Fig. 3.8.4.   Intersection of rows of a band matrix.

sets of i's. The first shaded area which is associated with i_1 (where $1 \leqslant i_1 \leqslant \beta + 1$) increases with i_1; the same is true for the shaded area associated with i_2, but the shaded area corresponding to i_3 remains the same as i_3 increases from $2\beta + 1$ to $n - 2\beta$. Furthermore, all areas associated with i_1 are less than those associated with i_2 which in turn are less than the area corresponding to i_3. Therefore, if we tacitly assume that the nondiagonal ones in B within the band are randomly distributed, then the values of v_is will generally increase as i increases from 1 to $2\beta + 1$ and after that remain the same until i equals $n - 2\beta$. From the symmetry of B, we can easily conclude that the situation is reversed for the bottom right-hand corner of B; in other words, v_i generally decreases as i increases from $n - 2\beta + 1$ to n. We also note that any two rows of B that are more than $2\beta + 1$ apart will have a zero intersection. All the above facts can be used to permute an arbitrary symmetric sparse matrix into a band form as follows.

Compute the usual (not Boolean) square of B and call it $\tilde{W}$. Then the (i, j) element of $\tilde{W}$ is the length of the intersection of row i with row j and v_i the sum of the lengths of the intersections of row i with rows of B is given by

$$(3.8.4) \qquad\qquad v_i = e_i'\tilde{W}V,$$

where V is an nth order column vector of all ones. We can now describe the method in the following algorithmic form (Tewarson, 1971).

1. Compute the usual square of B and denote it by $\tilde{W}$ and then compute v_i according to (3.8.4).

2. Compute

$$\beta = \max\left(\frac{\tau - n}{2n}, \frac{\rho_{max} - 1}{2}\right),$$

where τ is the total number of ones in B, ρ_{max} is the maximum number of ones in any one row of B. (Remark: This value of β is an underestimate, $(\tau - n)/2n$ is based on the assumption that the band is full and $(\rho_{max} - 1)/2$ on the assumption that the row with maximum number of elements can be permuted to be one of the rows i_3 of Fig. 3.8.4 that has no zeros within the band.)

3. Arrange the v_is in ascending order of their magnitudes. Denote the set of indices corresponding to the first 2β values of the v_is by CI (Corner Indices of band matrix) and the rest of indices by MI (Middle Indices). Divide the set CI into two subsets NW and SE (northwest and southeast corners of the band matrix) as follows (see Fig. 3.8.5). Determine $v_p = \min_i v_i$, $i \in$ CI; if there are ties, choose the v_p with minimum p. Assign p and all i in CI with $e_p'\tilde{W}e_i \neq 0$ to the set NW. Assign all indices j in CI to NW for which $e_i'\tilde{W}e_j \neq 0$ and $i \in$ NW. Repeat this procedure until no more indices of CI can be assigned to NW. In other words, $e_i'\tilde{W}e_q = 0$, for all i in NW and $q \notin$ NW but in CI. This determines all the indices for the NW corner. In the same manner determine

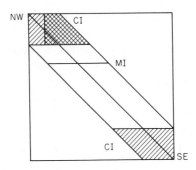

Fig. 3.8.5. Assignment of a row to a corner.

the indices which belong to the SW corner out of the remaining indices in CI. Assign the indices of CI that could not be assigned to either the NW or the SE corner to the set MI. Arrange the rows (and columns) in NW and SE in ascending and descending values of v_is respectively. (Note that whatever row permutations are performed on B (and A), the same permutations should also be performed on columns of B (and A) to preserve symmetry.) Perform additional row (and column) permutations, if needed, to get the rows in NW and SE into the shapes of the shaded regions of Fig. 3.8.5. Record all these row (column) interchanges.

4. Assign the rows in MI one at a time to either NW or SW as follow (see Fig. 3.8.5): Let V be an nth order column vector such the $e_i'V = 1$, if $i \in$ NW and zero otherwise. Then compute $\hat{v}_p = \max_j (e_j' \tilde{W} V)$, for all $j \in$ MI. Assign p to set NW. Repeat this procedure until approximately half the rows in MI have been assigned to NW. Record the order in which the rows in MI are assigned to NW. In the same manner, assign rows to the set SE, and record the order in which they are assigned to SE. (Remark: The sum of the lengths of intersections of row $j \in$ MI with the rows in set NW is the doubly shaded region in the northwest corner of Fig. 3.8.5. Clearly this region is maximum for a row that is adjacent to the last row in NW.)

5. The row interchanges in Step 3 and the order of row assignment in Step 4 give the required permutation P such that $\hat{B} = PBP'$ is a band matrix.

There are several other desirable forms for Gaussian elimination but it is not our aim to describe each and every one of them in this chapter. However, the four forms, BDF, BTF, BNTF, and BF, described in Sections 3.5, 3.6, 3.7, and 3.8 are the basic parts of most of the other forms (some of which are discussed in the next section.)

3.9. Other Desirable Forms

In Fig. 3.9.1 we have shown some of the other forms to which a given matrix can be permuted (Tewarson, 1971). The block diagonal form is part of both the *Singly Bordered Block Diagonal Form* (SBBDF) and the *Doubly Bordered Block Diagonal Form* (DBBDF). The BTF and BNTF are parts of the *Bordered Block Triangular* (BBTF) and the *Bordered Band Triangular Forms* (BBNTF) respectively. The band matrix is associated with the *Singly Bordered Band Form* (SBBF) and the *Doubly Bordered Band Form* (DBBF). Note that combinations of the various forms given in Fig. 3.9.1 are also possible; two of these are given in

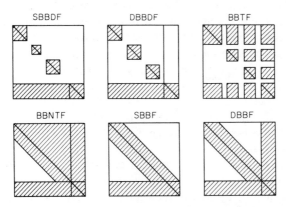

Fig. 3.9.1. Some simple desirable forms.

Fig. 3.9.2 (Jennings, 1966, 1968). Matrices of the second type result by renumbering tower frame stiffness matrices.

If *B* is SBBDF, then in the corresponding row graph the vertices associated with the rows in the border have generally a higher degree

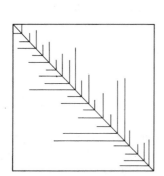 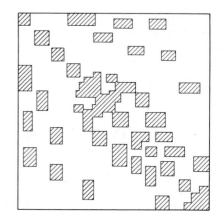

Fig. 3.9.2. Two other desirable forms.

than the other vertices. Furthermore, the removal of these vertices breaks the row graph into several disjoint subgraphs, each corresponding to a particular diagonal block of B. In the literature on graph theory, the *attachment set* is the subset of vertices; the removal of such a set (and all the edges associated with it) breaks the graph into two or more disjoint subgraphs (Mayoh, 1965). In power systems, the coupling transformers correspond to the points of attachment (Reid 1971, p. 125). If B is DBBDF then the vertices corresponding to the rows and columns associated with the borders constitute the "attachment set" of the graph of B. Let v_i be defined as in (3.8.4), then in the case of SBBDF and SBBF, we observe that all rows lying in the border have generally a larger v_i than others (since v_i is the sum of the lengths of intersections of the ith row with all the rows of B). This fact can be used to determine the rows which should belong to the borders. In the case of DBBDF and DBBF, it is also easy to determine the rows and columns belonging to the attachment set because they have generally larger v_is than the other rows and columns (Ogbuobiri *et al.*, 1970).

In the case of BBTF and BBNTF, if the rows in the border are moved to the top, then we get slightly different forms of BTF and BNTF, respectively (see Fig. 3.9.3). It is evident that we cannot use the methods of Section 3.6 to permute a given matrix into the modified BDF given in Fig. 3.9.3. However, the second method given in Section 3.7 can be used

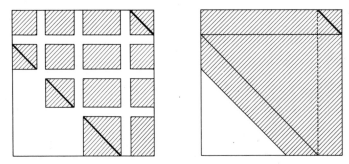

Fig. 3.9.3. Modified forms of BTF and BNTF.

to permute the given matrix to the band triangular form of Fig. 3.9.3. This can be followed (whenever possible) by further permutations according to Theorem 3.7.13 to get as many corner blocks as possible, since the first matrix in Fig. 3.9.3 can be thought of as BNTF with corner blocks.

We have described some simple practical methods for dealing with bordered forms. They work reasonably well for matrices which can be permuted to these forms with the nonzero elements evenly distributed in the shaded areas. At present, there is no definitive algorithm (like the ones given in Sections 3.5 and 3.6) for the determination of rows and/or columns which belong to the border or borders (Harary, 1971b), where only a few nonzero elements constitute the border or borders. One would like to have a simple, practical method for the determination of a smallest set of vertices of a graph (or digraph) whose removal makes it less connected.

We will now describe Steward's method (1969) of tearing (or removal of arcs of a diagraph to break cycles). It is especially useful in cases where the removal of a few arcs breaks large directed cycles and the corresponding diagonal blocks of the BTF matrix then become smaller (see Section 3.6). This method is practical only for tearing small blocks. We shall describe it by an example. Consider the matrix and its digraph given in Fig. 3.9.4. We determine one of the longest cycle [1, 4, 3, 6, 5, 1]. Two paths between the same vertices, oriented in the same direction, having no arcs in common and containing no cycles are said to be *parallel*. A path parallel to an arc in the long cycle is a *shunt*. In Fig. 3.9.4, four shunts are labeled $a, b, c,$ and d. The *order* of a shunt is the

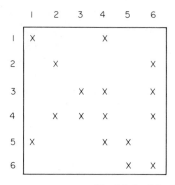
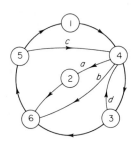

Fig. 3.9.4.   Matrix and its digraph.

length of the shunted path in the long cycle minus the length of the shunt; for example, c is [5, 4] and [5, 1, 4] is the path in the long cycle it shortens, hence its order is $2 - 1 = 1$. If an arc in the long cycle having a shunt is torn, then there remains a cycle path via the shunt which is the length of the long cycle minus the order of the shunt. Thus we tear the long cycle where it does not have too many shunts, since the shunts must also be torn. In Fig. 3.9.5, the shunt diagram associated with the long

Order		1	4	3	6	5	
0			B —— 2 —— E			a
1			B ———————— E			b
1			———————— E		B—— c
3			———————— E	B———————— d

Fig. 3.9.5.   Shunt diagram of long cycle.

cycle of Fig. 3.9.4 is given. B denotes the beginning and E the end of a shunt. It is evident that if we tear the arc [6, 5] then only the shunt d needs to be torn and therefore it is the best choice for tearing. In the digraph of Fig. 3.9.4, if the arcs [6, 5] and the shunt [3, 4] are removed (torn), then vertex 5 becomes an emitter and is relabeled as the first vertex. Now removing arcs [5, 1] and [5, 4], vertex 1 becomes an emitter, and is relabeled as the second vertex, and so on. The matrix associated with the relabeled digraph is given in Fig. 3.9.6. It is in BTF except for the torn element.

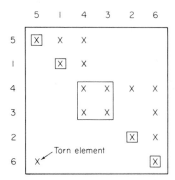

Fig. 3.9.6.   The matrix associated with the relabeled digraph.

The matrix in Fig. 3.9.6 is a special case of a BBTF matrix, with the border consisting of a single off-diagonal element. When there are several elements to be torn, all rows and columns of the matrix that contain the torn elements can be permuted to become its last rows and columns, such that the resulting matrix is in a BBTF. We shall now show how the EFIs for BTF and BBTF can be obtained.

3.10.   Inverses of BTF and BBTF

Consider the BBTF given by (3.3.1). Let all $A_{ii}, i = 1, 2, \ldots, p$ be nonsingular matrices. The method can be described in the following algorithmic form (Tewarson, 1972; Duff, 1972).

1. For $i = p - 1, p - 2, \ldots, 2, 1$ perform the following steps.

    (i) Transform A_{ii} to a unit upper triangular matrix U_{ii} and A_{pi} to a zero matrix by using the forward course of the *Gaussian elimination* (GE). This will lead to fill-in in A_{ip} and A_{pp}.

    (ii) Use the back substitution part of GE to transform U_{ii} to I and A_{ji} to zero for all $j < i$. This leads to fill-in in all A_{jp}s, $j \leqslant i$.

2. Use the forward course of GE to transform the modified A_{pp} to an upper triangular form U_{pp} and then back substitution to reduce U_{pp} to I and all the modified A_{jp}s, $j \neq p$ to zero.

It is evident that in this method no fill-in will take place in the matrices A_{ji}, $j < i$ and $i \neq p$.

If the given matrix is of BTF then for each i, $i = p, p - 1, \ldots, 2, 1$, we perform the following two steps:

(i) Transform A_{ii} to U_{ii}.

(ii) Reduce U_{ii} to I and A_{ji} to 0 for $j < i$.

Clearly, in this case there will be no fill-in in A_{ji}, $j \neq i$. Both of the above modifications of the GE are easy to incorporate and the corresponding EFIs are sparse. In both of the above methods, when transforming A_{ii} to U_{ii}, the various methods for reducing the fill-in described in this and the previous chapter can be utilized.

3.11.   Bibliography and Comments

The graph theoretic interpretation of the various desirable forms discussed in this chapter is given in Harary (1971a, b). A method which is analogous to the first method in Section 3.7 and minimizes $\sum_i \phi_i$ instead of $\max_i \phi_i$ is given in Tewarson (1967c). Alway and Martin (1965) have a program which does an educated search of possible permutations to determine the one for permuting a matrix to BF. Computer programs for the first and third methods of Section 3.8 are given in Rosen (1968) and Akyus and Utku (1968). Graph theoretic methods for Gaussian elimination are given in Parter (1961) and Rose (1970b). The equivalence of the Gaussian elimination to node elimination is shown in Ogbuobiri et al. (1970). Parter (1960) and Marimont (1969) make use of tearing. Graph theoretic methods for analyzing the structures and partitioning of matrices are given in Wenke (1964), Eufinger et al. (1968), Eufinger (1970). A good storage scheme for matrices of the forms given in Fig. 3.9.2 is described by Jennings (1966, 1968). An algorithm for solving symmetric positive definite band form matrices is given in Cantin (1971). Another algorithm for symmetric matrices is given in Jensen and Parks (1970). Partitioning and block elimination are discussed in George (1972), and Rose and Bunch (1972). To find the "points of attachment," it is also possible to use the dissection theory (Baty and Stewart, 1971), which is also called the "decomposition theory" in linear programming (Dantzig and Wolfe, 1961; Benders, 1962). It involves the determination of a nodal proximity matrix from the usual (not Boolean) powers of B. The elements of the nodal proximity matrix are then used to determine the attachment set.

4

Direct Triangular Decomposition

4.1. Introduction

In this chapter we will describe methods for expressing the given matrix A as the product of a lower triangular matrix $\tilde{L}$ with an upper triangular matrix U, namely,

(4.1.1) $$A = \tilde{L}U.$$

We shall be mainly concerned with the usefulness of these methods in the case of large sparse matrices. These methods are associated with the names of Crout, Doolittle, Choleski, Banachiewicz, and others (Westlake, 1968; Wilkinson, 1965).

If the factorization (4.1.1) is known, then it follows that

(4.1.2) $$A^{-1} = U^{-1}\tilde{L}^{-1}.$$

The factored forms of U^{-1} and $\tilde{L}^{-1}$ are trivial to determine as they are both triangular matrices. Therefore, we can use the above method in place of the one given in Chapter 2 for the determination of a factored form of inverse of A. If the solution of the system of equations $Ax = b$ is desired for only one right-hand side, then there is no need to evaluate (4.1.2). In this case, the solution x can be obtained by using back substitution to solve the systems

(4.1.3) $$\tilde{L}y = b$$

and

(4.1.4) $$Ux = y$$

for y and x, respectively.

The methods described in this chapter are essentially modifications of the Gaussian elimination described in Chapter 2. They have the advantage that the intermediate reduced matrices $A^{(k)}$ of Chapter 2 are not recorded, and furthermore if the inner products can be accumulated, then these methods are remarkably accurate.

4.2.   The Crout Method

Let the (i, j) elements of $\tilde{L}$ and U be denoted by l_{ij} and u_{ij}, respectively. Let us construct U as a unit upper triangular matrix, namely, $u_{kk} = 1$, $k = 1, 2, \ldots, n$. If we assume that the first $k - 1$ rows and columns of $\tilde{L}$ and U have already been determined (see Fig. 4.2.1), then in view of (4.1.1) and the facts that $l_{ip} = 0$ for $p > i$, $u_{pk} = 0$ for $p > k$, and $u_{kk} = 1$, we have

$$a_{ik} = l_{ik} + \sum_{p=1}^{k-1} l_{ip} u_{pk}, \qquad i \geq k,$$

which gives

(4.2.1) $$l_{ik} = a_{ik} - \sum_{p=1}^{k-1} l_{ip} u_{pk}, \qquad i \geq k.$$

Thus the kth column of L is now known. Now, from (4.1.1) and the fact that $l_{kp} = 0$ for $p > k$, we have

$$a_{kj} = l_{kk} u_{kj} + \sum_{p=1}^{k-1} l_{kp} u_{pj}, \qquad j > k,$$

which yields

$$(4.2.2) \qquad u_{kj} = \left(a_{kj} - \sum_{p=1}^{k-1} l_{kp} u_{pj} \right) \Big/ l_{kk}, \qquad j > k,$$

and the kth row of U is now known. Thus we have seen that if the first $k - 1$ rows of U and the first $k - 1$ columns of $\tilde{L}$ are known, then the kth row of U and the kth column of $\tilde{L}$ can be easily computed. The first column of $\tilde{L}$ is given by

$$(4.2.3) \qquad l_{i1} = a_{i1}, \qquad i = 1, 2, \ldots, n;$$

this follows from (4.1.1) and the fact that the first column of U is e_1. The first row of U is also easy to find from (4.1.1) and the fact that the first row of $\tilde{L}$ is $l_{11} e_1'$, namely,

$$(4.2.4) \qquad u_{1j} = a_{1j}/l_{11}, \qquad j > 1.$$

If we assume that for $k = 1$, $\sum_{p=1}^{k-1} (\cdots) = 0$, then (4.2.3) and (4.2.4) are particular cases of (4.2.1) and (4.2.2), respectively. In view of the above facts, all the rows and columns of $\tilde{L}$ and U can be easily determined.

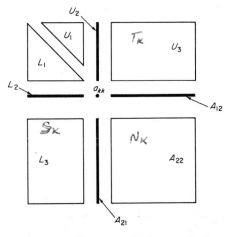

Fig. 4.2.1. Storage for the Crout method.

Fig. 4.2.1 illustrates how the various matrices are stored at the beginning of the kth step of the Crout method. The first $k - 1$ rows of

U are $[I + U_1, U_2, U_3]$ and the first $k - 1$ columns of $\tilde{L}$ are

$$\begin{bmatrix} L_1 \\ L_2 \\ L_3 \end{bmatrix}.$$

Note that the diagonal of U, which is I_n, is not stored. At the kth stage, we first use (4.2.1) to transform a_{kk} and A_{21} into the nontrivial elements of the kth column of $\tilde{L}$, namely,

$$(4.2.5) \qquad \begin{bmatrix} \hat{a}_{kk} \\ \hat{A}_{21} \end{bmatrix} = \begin{bmatrix} a_{kk} \\ A_{21} \end{bmatrix} - \begin{bmatrix} L_2 \\ L_3 \end{bmatrix} U_2,$$

where $l_{kk} = \hat{a}_{kk}$ and $l_{i+k,k} = e_i' \hat{A}_{21}$. This is followed by (4.2.2) to compute the nontrivial elements of the kth row of U, namely,

$$(4.2.6) \qquad \hat{A}_{12} = (A_{12} - L_2 U_3)/\hat{a}_{kk},$$

where $u_{k,j+k} = \hat{A}_{12} e_j$. To determine $\tilde{L}$ and U completely, the order of computation is as follows: first column of $\tilde{L}$, first row of U, second column of $\tilde{L}$, second row of U, and so forth. Once $\tilde{L}$ and U are known, it is easy to make use of (4.1.3) and (4.1.4) to evaluate y and x as follows:

$$(4.2.7) \qquad y_k = \left(b_k - \sum_{p=1}^{k-1} l_{kp} y_p \right) \Big/ l_{kk}, \qquad k = 1, 2, \ldots, n$$

and

$$(4.2.8) \qquad x_k = y_k - \sum_{p=k+1}^{n} u_{kp} x_p, \qquad k = n, n-1, \ldots, 1,$$

where $\displaystyle\sum_{p=1}^{0} (\cdots)$ and $\displaystyle\sum_{p=n+1}^{n} (\cdots)$ are both considered as zeros.

Since (4.2.7) and (4.2.2) are essentially the same, the column vector y can be generated during the Crout reduction by applying (4.2.2) to the augmented matrix $(A|b)$ instead of only to A and letting $a_{k,n+1} = b_k$ and $u_{p,n+1} = y_p$. The accuracy of the Crout method can be improved by partial pivoting as follows (Wilkinson, 1965). We compute l_{ik} according to (4.2.1), then determine

$$(4.2.9) \qquad |l_{sk}| = \max_i |l_{ik}|, \qquad i \geqslant k,$$

and interchange the sth and kth rows of $\tilde{L}$ and A. This is followed by the use of (4.2.1) for $i > k$ (to compute the rest of the elements of the kth column of $\tilde{L}$) and (4.2.2). A record of all such row interchanges is kept which yields a permutation matrix P such that

$$PA = \hat{L}\hat{U}$$

(where $\hat{L}$ and $\hat{U}$ are lower and unit upper triangular matrices). The above equation implies that

(4.2.10) $$A^{-1} = \hat{U}^{-1}\hat{L}^{-1}P.$$

Comparing (4.2.10) with (4.1.2), it is evident that whatever row permutations were applied during the triangular decomposition, the same permutations have to be applied to the columns of $\hat{U}^{-1}\hat{L}^{-1}$ to get the inverse of the given matrix A. Wilkinson (1965) has proved that triangular decomposition with partial pivoting is remarkably accurate if the inner products in (4.2.1) and (4.2.2) can be accumulated.

We notice that if $l_{kk} = 0$ then (4.2.2) cannot be evaluated. But this does not happen if (4.2.9) is used. For in this case l_{sk} cannot be zero as this would imply that the kth column of the lower triangular factor of A is zero and A is singular.

The relation between the Gaussian elimination and the direct triangular decomposition of Crout is as follows (Wilkinson, 1965). For nonsingular A, $\tilde{L}$, and U are unique if they exist and one of them is unit triangular. Therefore, from (2.2.7) and (4.1.1), it follows that $\tilde{L} = L^{-1}$ and the Us are identical in both equations. If identical row and column permutations are performed in both the Gaussian elimination and the Crout method, then once again the methods have the above-mentioned relationship.

4.3. Minimizing the Fill-in for the Crout Method

At the beginning of the kth stage of the Crout method it is possible to choose a nonzero element from the last $n - k + 1$ rows and columns of A and move it to the (k, k) position, such that the fill-in is minimized. In order to determine such an element, we shall need the following (Tewarson, 1969a). Let B_k denote the matrix which results when all the nonzero elements in the matrix of Fig. 4.2.1 are replaced by unity. Let the (i, j) elements of B_k be denoted by $b_{ij}^{(k)}$. Let S_k, T_k, and N_k denote the submatrices $b_{ij}^{(k)}, i \geqslant k, j < k$; $b_{ij}^{(k)}, i < k, j \geqslant k$; and $b_{ij}^{(k)}, i, j \geqslant k$. We define

$$(4.3.1) \qquad \Lambda_k = S_k * T_k,$$

where $*$ denotes Boolean matrix multiplication, in other words, the usual matrix multiplication with $1 + 1 = 1$. Let $\bar{\Lambda}_k$ be the matrix obtained from Λ_k by changing each of its nonzero elements to unity and vice versa, and

$$(4.3.2) \qquad \Delta_k = \bar{\Lambda}_k \oplus N_k,$$

where $\oplus$ denotes Boolean matrix addition, namely, $1 \oplus 1 = 1$. If V is a column vector of $n - k + 1$ ones, then we define

$$(4.3.3) \qquad \bar{c}^{(k)} = V' \Delta_k$$

and

$$(4.3.4) \qquad \bar{r}^{(k)} = \Delta_k V.$$

If the $\bar{r}_\alpha^{(k)}$ and $\bar{c}_\beta^{(k)}$ denote the αth and βth elements of $\bar{r}^{(k)}$ and $\bar{c}^{(k)}$, respectively, then we have the following theorem:

$(4.3.5)$ *THEOREM*   If $\bar{r}_s^{(k)} + \bar{c}_t^{(k)} = \max_{\alpha, \beta} (\bar{r}_\alpha^{(k)} + \bar{c}_\beta^{(k)})$, and complete cancellation in computing the inner products in (4.2.1) and (4.2.2) is neglected, then moving $a_{s+k-1, t+k-1}$ to the (k, k) position at the beginning of the kth step of the Crout method leads to the least local fill-in.

Proof   If $e_\alpha' \Delta_k e_\beta = 1$, then from (4.3.2) it follows that $e_\alpha' \bar{\Lambda}_k e_\beta$ and $e_\alpha' N_k e_\beta$ cannot both be zero. In view of the fact that $\bar{\Lambda}_k$ was obtained from Λ_k by changing its nonzeros to zeros and vice versa and equation (4.3.1), it follows that the equations $e_\alpha'(S_k * T_k)e_\beta = 1$ and $e_\alpha' N_k e_\beta = 0$ cannot both be true if $e_\alpha' \Delta_k e_\beta = 1$. If we interchange the kth and $(\beta + k - 1)$th columns of the matrix of Fig. 4.2.1 before the kth step of the Crout method, then from the definition of S_k, T_k, and N_k and the fact that we neglect the cancellation in computing the inner products, we have

$$\sum_{p=1}^{k-1} l_{ip}u_{pk} \neq 0 \Leftrightarrow e_\alpha'(S_k * T_k)e_\beta = 1,$$

where $\alpha + k - 1 = i$, and the element in the (i, k) position of the permuted matrix is zero, iff

$$e_\alpha'N_k e_\beta = 0.$$

If $e_\alpha'N_k e_\beta = 0$ and $e_\alpha'(S_k * T_k)e_\beta = 1$ are both satisfied, then according to (4.2.1), fill-in will take place, because a zero in the (i, k) position will become a nonzero. Thus we have seen that if $e_\alpha'\Delta_k e_\beta = 1$, and columns $\beta + k - 1$ and k are interchanged before the kth step of the Crout method, then no fill-in can take place in the (i, k) position. The total number of such places in which no fill-in can take place is $V'\Delta_k e_\beta$ which, in view of (4.3.3), is equal to $\bar{c}_\beta^{(k)}$. In exactly the same manner we can show that $\bar{r}_\alpha^{(k)}$ gives the total number of positions in which no fill-in takes place if rows k and $\alpha + k - 1$ are interchanged before the kth step of the Crout method. Therefore the least local fill-in will take place if the kth and $(s + k - 1)$th rows and the kth and $(t + k - 1)$th columns are interchanged before the kth step of the Crout method, where $\bar{r}_s^{(k)} + \bar{c}_t^{(k)} = \max_{\alpha,\beta} (\bar{r}_\alpha^{(k)} + \bar{c}_\beta^{(k)})$. This completes the proof of the theorem.

It is important that the element $a_{\hat{s}\hat{t}}$, where $\hat{s} = s + k - 1$ and $\hat{t} = t + k - 1$, chosen according to the above theorem does not become smaller than some pivot tolerance ε after it is modified according to (4.2.1), namely,

(4.3.6)
$$\left| a_{\hat{s}\hat{t}} - \sum_{p=1}^{k-1} l_{\hat{s}p}u_{p\hat{t}} \right| > \varepsilon.$$

The choice of ε has already been discussed in Section 2.3. It is generally not possible to make the test (4.3.6) prior to the use of Theorem 4.3.5, as this would involve a large amount of computation, and would be analogous to "complete pivoting" in the Crout method, which is not recommended (Wilkinson, 1965). In practice, a few of the elements which lead to the minimum or near minimum fill-in are first selected and then tested according to (4.3.6) before one of them is finally selected as the *pivot* for the kth step.

It is possible to use Theorem 4.3.5 by simulating the Crout method if we start with B_1 (the matrix obtained from A by replacing each of its nonzero elements by unity) and use Boolean multiplications and additions in (4.2.1) and (4.2.2) to record the fill-in. In this way the pivots for each stage are selected "a priori" and A can be permuted to have all such pivots on the leading diagonal before the actual Crout method is performed. This can be done if none of the computed pivots are smaller than ε. If A is positive definite then the pivots can be chosen in the above manner (Tinney and Walker, 1967; Gustavson *et al.*, 1970). Positive-definitive matrices and symmetric matrices which may not be positive definite are discussed in Section 4.5.

4.4.   The Doolittle (Black) Method

This is a variation of the Crout method in which, at the kth stage, only the kth rows of $\tilde{L}$ and U are generated (Westlake, 1968; Tinney and Walker, 1967). This is advantageous if A is stored by rows. In this method the elements of the kth row of $\tilde{L}$ are computed from left to right by using the formula

(4.4.1) $$l_{kj} = a_{kj} - \sum_{p=1}^{j-1} l_{kp}u_{pj}, \qquad j = 1, 2, \ldots, k$$

and the elements of the kth row of U are computed as in the Crout method by using (4.2.2).

The order of computation is: first row of $\tilde{L}$, first row of U, second row of $\tilde{L}$, second row of U, and so forth.

To minimize the local fill-in in this method it is not possible to use Theorem 4.3.5, as only the first row of S_k is known at the beginning of the kth step. It is perhaps best to simulate the Crout method and then choose the pivots a priori as mentioned in the previous section.

It is possible to generate the $\tilde{L}$, U matrices also by columns instead of rows by computing

$$(4.4.2) \quad u_{ik} = \left(a_{ik} - \sum_{p=1}^{i-1} l_{ip}u_{pk}\right)\bigg/ l_{ii}, \qquad i = 1, 2, \ldots, k-1$$

and then using (4.2.1) to compute l_{ik}, $i \geqslant k$. This variation of the Crout method is useful when nonzero elements of A are stored by columns.

In the next section we discuss the $\tilde{L}U$ factorization for symmetric matrices.

4.5. The Cholesky (Square-Root, Banachiewicz) Method

If the matrix A is nonsingular and symmetric, then it is well known that $A = \tilde{L}\tilde{L}'$ or $U'U$, provided that A has been arranged so that none of its northwest principal submatrices are singular. $\tilde{L}$ is a unique lower triangular and U a unique upper triangular matrix. If $A = U'U$, then

$$(4.5.1) \quad \sum_{p=1}^{k} u_{pk}u_{pj} = a_{kj}, \quad k \leqslant j.$$

and therefore the kth row of U is given by

$$(4.5.2) \quad u_{kk} = \left(a_{kk} - \sum_{p=1}^{k-1} u_{pk}^2\right)^{1/2}$$

and

$$(4.5.3) \quad u_{kj} = \left(a_{kj} - \sum_{p=1}^{k-1} u_{pk}u_{pj}\right)\bigg/ u_{kk}, \qquad j > k,$$

where we assume that $\sum_{p=1}^{k-1} (\cdots) = 0$ for $k = 1$. The above equations are used alternately to compute the rows of U. The diagonal elements u_{kk} may be complex numbers, if A is not positive definite.

If A is a positive-definite symmetric matrix, then the Cholesky method is the best method for triangular decomposition (Wilkinson, 1965). No row or column interchanges are required to keep round-off errors small. On the other hand, if A is symmetric but not positive definite, then pivoting has to be used to keep the round-off errors reasonable, and this destroys the symmetry. Taking the largest diagonal element at each stage preserves the symmetry but does not guarantee stability in terms of round-off errors.

In view of the above facts the Cholesky decomposition can be used for sparse symmetric matrices if arbitrary order of decomposition does not adversely affect numerical accuracy. In many practical applications this is fortunately true (for example, see Tinney and Walker, 1967).

In order to minimize the fill-in in the Cholesky method, we can use Theorem 4.3.5. However, in this case

$$(4.5.4) \qquad\qquad S_k = T_k',$$

which, in view of (4.3.1), (4.3.2), clearly implies that Δ_k is symmetric. Therefore, from (4.3.3) and (4.3.4), we can conclude that

$$(4.5.5) \qquad\qquad \bar{c}^{(k)} = (\bar{r}^{(k)})'.$$

In view of the above facts, instead of Theorem 4.3.5 the following corollary to it can be used.

(**4.5.6**)   *COROLLARY*   If $\bar{r}_s^{(k)} = \max_\alpha (\bar{r}_\alpha^{(k)})$, and complete cancellation in computing the inner products in (4.5.3) is neglected, then moving the $a_{s+k-1,s+k-1}$ to the (k, k) position at the beginning of the kth step of the Cholesky method leads to the least local fill-in.

Proof At any stage of the Cholesky method only the diagonal elements of A can be interchanged with each other, otherwise the symmetry will be destroyed. Therefore, in Theorem 4.3.5 we must take $\alpha = \beta$, and in view of (4.5.5), it follows that

$$\bar{r}_s^{(k)} + \bar{c}_s^{(k)} = \max_\alpha (\bar{r}_\alpha^{(k)} + \bar{c}_\alpha^{(k)}) \Leftrightarrow \bar{r}_s^{(k)} = \max_\alpha \bar{r}^{(k)}.$$

From this point on the proof of the corollary is the same as the proof of Theorem 4.3.5 (with l_{kp} replaced by u_{pk}).

It is possible to use the above corollary by simulating the Cholesky method if we start with the upper triangular part of B_1 (the matrix obtained from A by replacing each of its nonzero elements by unity) and use Boolean multiplications and additions in (4.5.3) to record the fill-in. Note that there is no need to simulate (4.5.2) and furthermore, since the denominator in the Boolean simulation of (4.5.3) is always unity, there is no need for division. Therefore, a permutation matrix P can be determined and the diagonal elements of A arranged with the help of the equation $\hat{A} = PAP$. The actual Cholesky method is then used on $\hat{A}$.

4.6. Desirable Forms for Triangular Decomposition

In Chapter 3, we gave some desirable forms into which a given matrix can be permuted a priori so that the fill-in, if any, is limited to certain regions of these forms.

We will now show that the desirable forms for triangular decomposition are also the same. At the end of Section 4.2, we pointed out the fact that $\tilde{L} = L^{-1}$ where $\tilde{L}$ and L are the lower triangular matrices obtained by the Crout and the Gaussian elimination methods, respectively, provided that the order and the choice of pivots are identical in both cases. Now, in view of (2.2.6), we have

$$L_n \cdots L_2 L_1 \tilde{L} = I_n,$$

and from (2.2.3) and (2.2.4) it is evident that L_k reduces the kth column of $L_{k-1} \cdots L_2 L_1 \tilde{L}$ to e_k and leaves all the other columns unchanged, thus we conclude that

(4.6.1)
$$\eta_i^{(k)} = 0, \qquad i < k;$$
$$\eta_k^{(k)} = 1/l_{ii} \quad \text{and} \quad \eta_i^{(k)} = -l_{ik}/l_{kk}, \quad i > k.$$

In view of above facts, evidently the elements of $\tilde{L}$ and the nontrivial elements of the factors of L have the same pattern of nonzeros and therefore can be stored in identical storage locations. Thus the fill-in in both the Crout and the Gaussian elimination methods is identical, and therefore the same forms are desirable for both methods.

If A is a symmetric matrix or a matrix that is symmetric in the pattern of nonzero off-diagonal elements, then the band form (BF), doubly bordered band form (DBBF), block diagonal form (BDF), doubly bordered block diagonal form (DBBDF) or a combination of these forms are desirable for the Crout method (Cholesky method if $A = A'$).

If A is not symmetric, then it can be permuted to the singly bordered band form (SBBF), singly bordered block diagonal form (SBBDF), band triangular form (BNTF), bordered band triangular form (BBNTF), block triangular form (BTF), bordered block triangular form (BBTF), or a combination of these before the Crout method is used.

4.7. Bibliography and Comments

The problem of optimal ordering for triangular decomposition of matrices occurring in some of the practical applications is discussed by Carpentier (1963), Sato and Tinney (1963), Edelman (1963, 1968), Chang (1969), McCormick (1969), Tinney (1969), Ashkenazi (1971), and Jennings and Tuff (1971). In problems arising in many electrical network problems the triangular decomposition is especially useful (Tinney and Walker, 1967; Erisman, 1972). Some theoretical results comparing the Crout method with other methods are given in Brayton et al. (1969).

The Crout method can be generalized in the sense that, for any given k, we let either $u_{kk} = 1$ or $l_{kk} = 1$. In this case, the number of divisions can often be reduced if the batch having the smaller number of nonzeros is normalized (Gustavson et al., 1970).

The Gauss–Jordan Elimination

5.1. Introduction

If at each stage in the Gaussian elimination (see Section 2.2) not only the nonzero elements below the diagonal but also those above the diagonal are eliminated, then the process is called the *Gauss–Jordan elimination*. Thus the given coefficient matrix A is reduced directly to the identity matrix; this is in contrast with the Gaussian elimination, where A is first transformed to an upper (unit) triangular matrix U which is later reduced to the identity matrix.

In this chapter we shall describe the Gauss–Jordan elimination and show how the matrices associated with the various stages of the elimination process are utilized to express A^{-1} in a factored form which is called the *Product Form of Inverse* (PFI). We shall also show how a sparse PFI of a given sparse matrix can be obtained.

5.2.   The Basic Method

In the Gauss–Jordan elimination (GJE) a sequence of elementary transformations (row operations) is applied to the given matrix A to reduce it to the identity matrix I_n. The same sequence of transformations, when applied to the right-hand side b of the system $Ax = b$, yields the solution x (Faddeev and Faddeeva, 1963; Fox, 1965).

Let $A^{(k)}$ denote the matrix at the start of the kth step of the elimination, where $k = 1, 2, \ldots, n$, and $A^{(1)} \equiv A$ and $A^{(n+1)} = I_n$. Let $a_{ij}^{(k)}$ be the (i, j) element of $A^{(k)}$. The matrix $A^{(k)}$ is identical to I_n in the first $k - 1$ columns. At the kth step, the kth column of $A^{(k)}$ is transformed to e_k by elementary row operations, namely,

(5.2.1) $A^{(k+1)} = T_k A^{(k)},$

where

(5.2.2) $T_k = I_n + (\zeta^{(k)} - e_k)e_k',$

and the elements of the column vector $\zeta^{(k)}$ are given by

(5.2.3) $\zeta_i^{(k)} = -a_{ik}^{(k)}/a_{kk}^{(k)}, \qquad i \neq k \qquad$ and $\qquad \zeta_k^{(k)} = 1/a_{kk}^{(k)}.$

Now, from (5.2.1) and the facts that $A^{(1)} \equiv A$ and $A^{(n+1)} = I_n$, we have

$$T_n \cdots T_2 T_1 A = I_n,$$

which gives the *Product Form of Inverse* (PFI) of A as

(5.2.4) $A^{-1} = T_n \cdots T_2 T_1.$

If at the end of the kth stage of the GJE, the $\zeta^{(k)}$ is stored in place of the kth column of $A^{(k+1)}$ (which is not needed later), then the nontrivial elements of PFI would replace the matrix A by the completion of the GJE.

The PFI is an essential part of most linear programming computer codes where minimizing the fill-in is an important consideration (Dantzig and Orchard-Hays, 1954; Larson, 1962; Smith and Orchard-Hays, 1963; Wolfe and Cutler, 1963; Tewarson, 1966, 1967a; Orchard-Hays, 1968; Dantzig et al., 1969; Brayton et al., 1969). We will discuss the minimization of the fill-in of nonzeros during the Gauss–Jordan elimination in Section 5.4.

In case the solutions of the system of linear equations $Ax = b$ are required for only a few right-hand sides, then there is no need to save the PFI, as each T_k can also be applied to the right-hand sides at the same time it is applied to $A^{(k)}$ in (5.2.1). Even in this case, minimizing the fill-in of nonzero elements, when $A^{(k)}$ is transformed in $A^{(k+1)}$ according to (5.2.1), is very useful for large sparse matrices.

In the next section we discuss the relation between the elimination form of inverse (EFI) which was defined in Section 2.4 and the PFI.

5.3.   The Relationship between the PFI and the EFI

We recall from Section 2.2 that during the back substitution part of the Gaussian elimination (GE), the factored form of inverse of the unit upper triangular matrix U is obtained by choosing its diagonal elements in sequence as pivots by starting from the bottom right-hand corner. In this case, no fill-in of nonzero elements can take place and the non-trivial elements of the factored form of the inverse of U are obtained by just changing the signs of those elements of U which lie above the diagonal. However, to compute U^{-1} in this manner, all the rows of U have to be known and one has to wait until the forward course of the GE is complete.

An alternative way to find U^{-1} is to choose the pivots by starting with the top left-hand corner and then proceed down the diagonal. In this case, some fill-in of nonzero elements will generally take place. However, at any particular stage k, only the first k rows of U are needed. Since at the completion of the kth stage of the forward course of the GE the first k rows of U are known, U^{-1} can be computed during the forward course of the GE. This is precisely what is done in the Gauss–Jordan elimination (GJE): The computation of U^{-1} in a factored form (which we will show to be the same as the explicit U^{-1}) is combined with the forward course of the GE. We will see that in the GJE, eliminating the elements above the diagonal is equivalent to computing U^{-1} by the alternative method given above. This method for computing the inverse of the unit upper triangular matrix U can be described mathematically as follows:

Let

(5.3.1) $$U^{(k+1)} = \tilde{U}_k U^{(k)}, \qquad k = 1, 2, \ldots, n,$$

where

(5.3.2) $$U^{(1)} = U, U^{(n+1)} = I_n$$

and

(5.3.3) $$\tilde{U}_k = I_n + \tilde{\xi}^{(k)} e_k'$$

with the elements of the column vector $\tilde{\xi}^{(k)}$ given by

(5.3.4) $\tilde{\xi}_i^{(k)} = 0, \quad i \geqslant k \qquad$ and $\qquad \tilde{\xi}_i^{(k)} = -u_{ik}^{(k)}, \qquad i < k$

[$u_{ij}^{(k)}$ is the (i, j) element of $U^{(k)}$].

In view of (5.3.3) and (5.3.4), we have $\tilde{U}_1 = I_n$ and therefore from (5.3.1) and (5.3.2), it follows that

(5.3.5) $$U^{-1} = \tilde{U}_n \cdots \tilde{U}_3 \tilde{U}_2.$$

We will need the following results to see the relationship between the GE and the GJE (Brayton *et al.*, 1969). Let L_k, T_k, and $\tilde{U}_k$ be defined by (2.2.3), (5.2.2), and (5.3.1), respectively, then we have

(5.3.6)   *LEMMA*   The last $n - k + 1$ rows of $L_k \cdots L_2 L_1 A$ and $T_k \cdots T_2 T_1 A$ are identical for $k = 1, 2, \ldots, n$.

Proof   The lemma is certainly true for $k = 1$, for both in the GE and the GJE the first column of A is reduced to e_1 by identical row operations taking $a_{11}^{(1)}$ as the pivot. Let us assume that the lemma holds for some k, then in the $(k + 1)$th stage of both the GE and the GJE, in the $(k + 1)$th column the $(k + 1)$th element is made equal to unity and all elements below it are made zero; in other words, identical row operations are performed on the last $n - k$ rows. Therefore, the last $n - k$ rows of $L_{k+1} \cdots L_2 L_1 A$ and $T_{k+1} \cdots T_2 T_1 A$ are also identical and, by induction, the proof of the lemma is complete.

(5.3.7)   *LEMMA*   The last $n - k + 1$ rows of L_k and T_k are identical for $k = 1, 2, \ldots, n$.

Proof In view of (2.2.4) and (5.2.3), we see that the nontrivial elements of the last $n - k + 1$ rows of both L_k and T_k are formed from the last $n - k + 1$ elements of the kth columns of $L_{k-1} \cdots L_2 L_1 A$ and $T_{k-1} \cdots T_2 T_1 A$, respectively. Therefore, in view of Lemma 5.3.6, the last $n - k + 1$ rows of L_k and T_k are identical.

(5.3.8)   *LEMMA*   If $U^{(k)}$ and $A^{(k)}$ are defined by (5.3.1) and (5.2.1), respectively, then for $k = 2, 3, \ldots, n$ the first $k - 1$ rows of both these matrices are identical.

Proof Since $e_1' L_1 A = e_1' U$, therefore, in view of Lemma 5.3.6 and the fact that $\tilde{U}_1 = I_n$, we have $e_1' A^{(2)} = e_1' T_1 A = e_1' L_1 A = e_1' U = e_1' \tilde{U}_1 U = e_1' U^{(2)}$; thus the lemma is true for $k = 2$. Suppose that the lemma is true for some k, then from (5.2.1), Lemma 5.3.6, and the fact that at the end of the kth step of the forward course of the GE $e_k' L_k \cdots L_2 L_1 A = e_k' U$, we have

$$\begin{aligned} e_k' A^{(k+1)} &= e_k' T_k \cdots T_2 T_1 A \\ &= e_k' L_k \cdots L_2 L_1 A = e_k' U \\ &= e_k' U^{(k+1)}. \end{aligned}$$

Now from (5.2.1), (5.2.2), (5.2.3), (5.3.1), (5.3.3), and (5.3.4), it follows that identical row operations are performed during the kth stages on the first $k - 1$ rows of both the matrices $A^{(k)}$ and $U^{(k)}$. Therefore, not only the kth rows but also the first $k - 1$ rows of $A^{(k+1)}$ and $U^{(k+1)}$ are identical. Induction on k completes the proof of the lemma.

(5.3.9)   *LEMMA*   The first $k - 1$ rows of $\tilde{U}_k$ and T_k are identical.

Proof In view of (5.3.3), (5.3.4), (5.2.2), and (5.2.3), we see that the nontrivial elements in the first $k - 1$ rows of both $\tilde{U}_k$ and T_k are formed from the first $k - 1$ elements of the kth columns of $U^{(k)}$ and $A^{(k)}$, respectively. Therefore, from Lemma

5.3.8, we conclude that the first $k - 1$ rows of $\tilde{U}_k$ and T_k are identical.

(5.3.10)   *LEMMA*   If $\tilde{U}_k$ and U^{-1} are given by (5.3.3) and (5.3.5), respectively, then the kth columns of these matrices are identical.

Proof   From (5.3.3) and (5.3.4) we have

$$\tilde{U}_p e_q = (I_n + \xi^{(p)} e_p')e_q = e_q, \qquad p \neq q,$$

$$\tilde{U}_p e_p = e_p + \xi^{(p)},$$

and

$$\tilde{U}_q \xi^{(p)} = (I_n + \xi^{(q)} e_q')\xi^{(p)}$$

$$= \xi^{(p)}, \qquad \text{since} \quad e_q' \xi^{(p)} = 0, \qquad q > p.$$

Therefore, we have

$$U^{-1} e_k = \tilde{U}_n \cdots \tilde{U}_2 e_k = \tilde{U}_n \cdots \tilde{U}_{k+1} \tilde{U}_k e_k$$

$$= \tilde{U}_n \cdots \tilde{U}_{k+1}(e_k + \xi^{(k)}) = e_k + \xi^{(k)} = \tilde{U}_k e_k,$$

which completes the proof.

   In view of Lemmas 5.3.7, 5.3.9, and 5.3.10, we have the following two equations which show the relationship between the PFI and EFI:

(5.3.11) $e_i' T_k e_k = e_i' \tilde{U}_k e_k = e_i' U^{-1} e_k, \qquad i < k$

and

(5.3.12) $e_i' T_k e_k = e_i' L_k e_k, \qquad i \geqslant k.$

From (5.3.11) it is clear that the nontrivial elements of the PFI which lie above the leading diagonal are the elements of the explicit form of U^{-1}, where U is the unit upper triangular form obtained at the end of the forward course of the GE. On the other hand, from (5.3.12), it follows that the nontrivial elements of both the PFI and the EFI which lie on or below the diagonal are identical. Thus the zero nonzero structure of both the PFI and the EFI is the same on or below the diagonal, but above it the PFI has the zero nonzero structure of U^{-1}

while the EFI has that of U. In general, the U^{-1} has more nonzeros than U and, therefore, the PFI is usually not as sparse as the EFI.

5.4. Minimizing the Total Number of Nonzeros in the PFI

In view of the close relationship between the PFI and the EFI, the discussion about pivoting and round-off errors given in Section 2.3 holds in the case of the PFI as well. The methods for minimizing the total number of nonzero elements in the PFI are essentially similar to those given in Section 2.5 (Markowitz, 1957; Larson, 1962; Smith and Orchard-Hays, 1963; Wolfe and Cutler, 1963; Dickson, 1965; Tewarson, 1966, 1967a,b; Orchard-Hays, 1968). We shall now briefly describe some of the relevant modifications that should be made in the methods given in Section 2.5, so that they can be used for the PFI.

At the kth stage of the GJE multiples of the kth row of $A^{(k)}$ are added to all the other rows and, as a result, fill-in of nonzero elements takes place not only below (as in the GE) but also above the diagonal. In order to minimize this fill-in, we need the following results.

If at the beginning of the kth step of the GJE we interchange the rows s and k and the columns t and k, then instead of $a_{kk}^{(k)}$, another element $a_{st}^{(k)}$ is taken as the pivot. Of course, $|a_{st}^{(k)}|$ has to be greater than the pivot tolerance ε. The effect of the above pivot choice is as follows: If P_k and Q_k are the matrices obtained by interchanging the sth and the kth rows and the tth and the kth columns of I_n, respectively, then

$$(5.4.1) \qquad \hat{A}^{(k)} = P_k A^{(k)} Q_k$$

has $a_{st}^{(k)}$ in the (k, k) position. Then, instead of (5.2.1) and (5.2.3), we have

$$(5.4.2) \qquad A^{(k+1)} = T_k \hat{A}^{(k)}, \qquad k = 1, 2, \ldots, n$$

and

$$(5.4.3) \quad \zeta_i^{(k)} = -\hat{a}_{ik}^{(k)}/\hat{a}_{kk}^{(k)}, \quad i \neq k \qquad \text{and} \qquad \zeta_k^{(k)} = 1/\hat{a}_{kk}^{(k)},$$

and from (5.4.1) and (5.4.2) we have

(5.4.4) $A^{-1} = Q_1 Q_2 \cdots Q_n T_n P_n \cdots T_2 P_2 T_1 P_1.$

Let B_k be the matrix obtained from the last $n - k + 1$ columns of $A^{(k)}$ by replacing all its nonzero elements by unity. (Note that the matrices $A^{(k)}$ and B_k are not identical to the corresponding matrices in Section 2.5). We now have the following theorem (Tewarson, 1967b) which is similar to Theorem 2.5.5.

(5.4.5)   *THEOREM*   If $a_{i,j+k-1}^{(k)}$, where $i \geqslant k$, is chosen as a pivot at the kth stage of the GJE, then the local fill-in is given by the (i, j) element of the matrix G_k with

(5.4.6) $G_k = B_k \bar{B}_k' B_k,$

where $\bar{B}_k'$ is the transpose of the matrix which results when each zero element of B_k is changed to unity and vice versa.

Proof   If in $A^{(k)}$ the $(p, q + k - 1)$ element is zero but both the $(i, q + k - 1)$ and the $(p, j + k - 1)$ elements are non-zero, then from (5.4.1), (5.4.2), (5.2.2), and (5.4.3), it follows that the $(p, q + k - 1)$ element in $A^{(k+1)}$ will be nonzero. From this point on the remaining steps in the proof of the present theorem are identical to those in Theorem 2.5.5, provided that the GE is replaced by the GJE and M is taken as an $n \times (n - k + 1)$ matrix of all ones. We are therefore not including these steps here.

The following corollary is a direct consequence of Theorem 5.4.5, and for this reason its proof is omitted.

(5.4.7)   *COROLLARY*   If at the kth stage of the GJE $a_{st}^{(k)}$ is chosen as the pivot, where $s \geqslant k, t = \beta + k - 1$, and s and β are given by

(5.4.8) $g_{s\beta}^{(k)} = \min_{i,j} e_i' G_k \tilde{e}_j,$

for all $|a_{i,j+k-1}^{(k)}| > \varepsilon,$ $i \geqslant k,$

(ε is some suitably chosen pivot tolerance and $\tilde{e}_j$ is the jth column of I_{n-k+1}), then the local fill-in will be minimized.

Since the elements of the T_ks are computed from the elements of the $A^{(k)}$s, minimizing the local fill-in in the $A^{(k)}$s will therefore minimize the number of nonzero elements on the PFI, provided that the local minima lead to a global minimum. This may be true for some matrices but is not so for arbitrary matrices.

A simpler, though less accurate method of finding a pivot which tends to keep the local fill-in small is based on the following theorem (Markowitz, 1957).

(5.4.9) *THEOREM*   If $a_{i,j+k-1}^{(k)}$, with $i \geqslant k$, is chosen as a pivot at the kth stage of the GJE, then the maximum possible fill-in (not the actual fill-in) is given by

(5.4.10)
$$\hat{g}_{ij}^{(k)} = (r_i^{(k)} - 1)(c_j^{(k)} - 1),$$

where

(5.4.11)
$$r_i^{(k)} = e_i' B_k V_k, \qquad c_j^{(k)} = V' B_k \tilde{e}_j,$$

V_k and V are column vectors with $n - k + 1$ and n ones, respectively, and $\tilde{e}_j$ is the jth column of I_{n-k+1}.

Proof From (5.4.11) it is evident that $r_i^{(k)}$ and $c_j^{(k)}$ denote, respectively, the total number of nonzero elements in the ith row and the $j + k - 1$ column of $A^{(k)}$. Therefore, the maximum possible fill-in that can take place when the $(i, j + k - 1)$ element of $A^{(k)}$ is chosen as pivot is equal to $(r_i^{(k)} - 1)(c_j^{(k)} - 1)$, which completes the proof.

It is easy to show that $\hat{g}_{ij}^{(k)}$ is the (i, j) element of the matrix $\hat{G}_k$ as follows: From (5.4.10), (5.4.11), and the fact that $e_i'V = V_k'\tilde{e}_j = 1$, we have

$$\hat{g}_{ij}^{(k)} = (e_i'B_k V_k - 1)(V'B_k\tilde{e}_j - 1)$$
$$= e_i'(B_k V_k - V)(V'B_k - V_k')\tilde{e}_j,$$

therefore,

(5.4.12)
$$\hat{G}_k = (B_k V_k - V)(V'B_k - V_k').$$

In order to make use of Theorem 5.4.9, instead of (5.4.8), we use the following equation to select the pivot at the kth stage

(5.4.13) $\hat{g}_{s\beta}^{(k)} = \min e_i'\hat{G}_k\tilde{e}_j, \qquad$ for all $|a_{i,j+k-1}^{(k)}| > \varepsilon.$

Note that the pivot $a_{s,\beta+k-1}^{(k)}$ chosen according to (5.4.13), does not necessarily lead to least local fill-in.

The methods for minimizing the storage of the EFI based on a priori column permutations given in Section 3.2, can also be applied to the PFI, due to the following reasons: For $k = 1$, equations (3.2.2) and (5.4.11) are identical and therefore, the $c_j^{(1)}$'s, $\gamma_j^{(1)}$'s, and $\lambda_j^{(1)}$'s will be identical for the GE and the GJE. Furthermore, in both the GJE and the GE, at the $(k + 1)$th stage, the pivots can only be chosen in the last $n - k$ rows and columns; therefore, only the last $n - k$ elements of the $r^{(k)}$'s associated with the GJE need be updated at the kth stage. This, along with the fact that B_k is an n by $n - k + 1$ matrix (not an $n - k + 1$ by $n - k + 1$ matrix as in the GE), permits us to restate Theorem 3.2.12.

(5.4.14) *THEOREM* If the nonzero elements in the last $n - k + 1$ rows and columns of $\hat{A}^{(k)}$ are randomly distributed in those rows and columns and, for $i \geqslant k$, $\hat{r}_i^{(k)}$ is the number of nonzero elements in the ith row of $\hat{A}^{(k)}$, then for $i > k$, $\tilde{r}_i^{(k+1)}$, the expected number of nonzero elements in the ith row of $A^{(k-1)}$, is given by

(5.4.15) $$\tilde{r}_i^{(k+1)} = \hat{r}_i^{(k)}, \qquad \hat{a}_{i,k}^{(k)} = 0$$

(5.4.16) $$\tilde{r}_i^{(k+1)} = \hat{r}_i^{(k)} + \hat{r}_k^{(k)} - 2$$

$$-\frac{(\hat{r}_i^{(k)} - 1)(\hat{r}_k^{(k)} - 1)}{n - k}, \qquad \hat{a}_{i,k}^{(k)} \neq 0$$

Proof The same as that of Theorem 3.2.12.

Thus we have seen that the columns of A can be arranged a priori according to the ascending values of the $c_j^{(1)}$'s, $\gamma_j^{(1)}$'s, or $\lambda_j^{(1)}$'s, and the pivots are chosen according to (3.2.11), but $\hat{r}^{(k)}$ is updated by using (5.4.15) and (5.4.16) instead of (3.2.13) and (3.2.14) (Tewarson, 1966, 1967a; Orchard-Hays, 1968).

As in Section 3.3, it is possible to apply row and column permutations P and Q to A such that $\hat{A} = PAQ$ is in one of the forms which are desirable for the GJE. This we shall now discuss.

5.5.   *Desirable Forms for the GJE*

If the pivots are taken sequentially on the diagonal of the matrix A starting from the top left-hand-side corner, then the following of the forms described in Chapter 3 are also desirable for the GJE: BDF, SBBDF, DBBDF, transpose of the BTF (lower block triangular form) and transpose of the BBTF (bordered lower block triangular form). Methods for permuting A to one of these forms have already been given in Chapter 3.

A desirable form of $\hat{A}$, which is incorporated in some computer codes (Orchard-Hays, 1968), consists of finding permutation matrices P and Q such that

$$PAQ = \hat{A} = \begin{bmatrix} I & A_{12} & A_{13} & A_{14} \\ 0 & A_{22} & 0 & 0 \\ 0 & A_{32} & A_{33} & 0 \\ 0 & A_{42} & A_{43} & A_{44} \end{bmatrix},$$

where A_{22} and A_{44} are lower triangular matrices and A_{33} is a square matrix. Pivots are chosen in I, A_{22}, A_{33}, and A_{44}, in that order. Except for A_{33}, all the pivots are chosen on the diagonal, and the only fill-in taking place is in the regions A_{13}, A_{33}, and A_{43} when the pivots are chosen in A_{33}. This fill-in can be minimized by any one of the methods given in Section 5.4.

5.6.   *Bibliography and Comments*

The GJE is described in most texts on numerical analysis, for example, Hildebrand (1956), Faddeev and Faddeeva (1963), Fox (1965), Ralston (1965), and Westlake (1968). The PFI is described in Dantzig and Orchard-Hays (1954), Gass (1958), Hadley (1962), and Dantzig (1963a).

Computational experiments in the use of the PFI and techniques for keeping it sparse are given in Larson (1962), Smith and Orchard-Hays

(1963), Wolfe and Cutler (1963), Dickson (1965), Baumann (1965), Tewarson (1966, 1967a), Orchard-Hays (1968), Dantzig *et al.* (1969), Brayton *et al.* (1969), Beale (1971), and de Buchet (1971).

A comparison of the EFI and the PFI in terms of fill-in for randomly generated sparse matrices is given by Brayton *et al.* (1969).

Orthogonalization Methods

6.1. Introduction

In many practical applications there is a need to transform a given sparse matrix into another matrix whose columns are orthonormal. The use of orthonormalizing codes is well known (Davis, 1962). In this chapter we will discuss the problem of finding the optimum order in which the columns of the given sparse matrix should be orthonormalized, such that the resulting matrix is as sparse as possible. We shall be concerned with the *Gram–Schmidt, the Householder, and the Givens methods* for orthonormalization (Tewarson, 1968a, 1970a).

6.2. The Gram–Schmidt Method

Let A denote an $m \times n$ matrix of rank n where $m \geq n$. The *Gram–Schmidt method* involves the determination of an upper triangular matrix $\hat{U}$ such that the columns of $A\hat{U}$ are orthonormal (Davis 1962, Rice 1966). If A is sparse, then it is generally advantageous to find a permutation matrix Q such that $AQ\hat{U}$ and $\hat{U}$ are both sparse. Methods for doing this will be discussed in the next section. In the present section, we shall describe a slightly modified version of the Gram–Schmidt method which is more accurate than the usual method in terms of round-off errors (Rice, 1966; Tewarson, 1968a). It is called the *Revised Gram–Schmidt (RGS) method* and consists of n steps, such that for $k = 1, \ldots, n$, if $A^{(k)}$ denotes the matrix at the beginning of the kth step, where $A^{(1)} \equiv A$, then, at the end of n steps, all the columns of $A^{(n+1)}$ are orthonormal. In the matrix $A^{(k)}$, the first $k - 1$ columns are orthonormal and the kth column is orthogonal to these. In the kth step, the kth column of $A^{(k)}$ is normalized and the last $n - k$ columns are made orthogonal to it; the resulting matrix is denoted by $A^{(k+1)}$. If $a_j^{(k)}$ and $a_{ij}^{(k)}$ denote, respectively, the jth column and the (i, j) element of $A^{(k)}$, then the RGS method can be described mathematically as follows:

(6.2.1) $$A^{(k+1)} = A^{(k)}\hat{U}_k, \qquad k = 1, 2, \ldots, n,$$

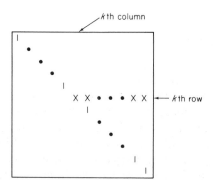

Fig. 6.2.1.   The elementary matrix $\hat{U}_k$ at the kth stage.

where the elementary upper triangular matrix $\hat{U}_k$ (see Fig. 6.2.1) is given by

(6.2.2) $$\hat{U}_k = I_n + e_k(\hat{\zeta}^{(k)} - e_k')$$

with the elements of the row vector $\hat{\xi}^{(k)}$ defined as follows:

(6.2.3)
$$\hat{\xi}_j^{(k)} = 0, \quad j < k;$$

$$\hat{\xi}_k^{(k)} = 1/(a_k^{(k)'} a_k^{(k)})^{1/2} \quad \text{and} \quad \hat{\xi}_j^{(k)} = -a_k^{(k)'} a_j^{(k)}/a_k^{(k)'} a_k^{(k)}, \quad j > k.$$

The $\hat{U}_k$ consists of ones on the diagonal except in the kth row which has $\hat{\xi}^{(k)}$'s on and to the right of the diagonal. The rest of the elements of $\hat{U}_k$ are zero. From (6.2.1), (6.2.2), and (6.2.3), it follows that

(6.2.4)
$$
\begin{aligned}
a_j^{(k+1)} &= a_j^{(k)}, & j < k \\
&= a_k^{(k)}/(a_k^{(k)'} a_k^{(k)})^{1/2}, & j = k \\
&= a_j^{(k)} - (a_j^{(k)'} a_k^{(k)}/a_k^{(k)'} a_k^{(k)})a_k^{(k)}, & j > k.
\end{aligned}
$$

This is the form in which the RGS is generally given in text books.

If A is a square matrix, namely, $m = n$, then from (6.2.1) and the fact that the inverse of a matrix with orthonormal columns is equal to its transpose, it follows that

(6.2.5)
$$A^{-1} = \hat{U}_1 \hat{U}_2 \cdots \hat{U}_n A^{(n+1)'}.$$

Thus the RGS method can be used for computing A^{-1}.

6.3. Minimizing the Nonzeros in the RGS Method

From (6.2.4), it is evident that in all the columns that have a nonzero inner product with $a_k^{(k)}$, some fill-in may take place. Therefore, in order to minimize the local fill-in, out of the last $n - k + 1$ columns a column $a_s^{(k)}$ is chosen which would lead to the least fill-in, if it is made the kth column prior to the kth step of the RGS method. This is done by interchanging the kth and the sth columns of $A^{(k)}$, namely,

(6.3.1)
$$\hat{A}^{(k)} = A^{(k)} Q_k,$$

where Q_k is obtained by interchanging the kth and the sth columns of

I_n. Then, instead of (6.2.1), we use

(6.3.2) $$A^{(k+1)} = \hat{A}^{(k)} \hat{U}_k, \qquad k = 1, 2, \ldots, n,$$

and if $\hat{a}_j^{(k)}$ and $\hat{a}_{ij}^{(k)}$ denote, respectively, the jth column and the (i, j) element of $\hat{A}^{(k)}$, then $\hat{U}_k$ is given by (6.2.2), but the elements of $\hat{\xi}^{(k)}$ are given by

(6.3.3)
$$\hat{\xi}_j^{(k)} = 0, \quad j < k;$$

$$\hat{\xi}_k^{(k)} = 1/(\hat{a}_k^{(k)'}\hat{a}_k^{(k)})^{1/2} \quad \text{and} \quad \hat{\xi}_j^{(k)} = -\hat{a}_k^{(k)'}\hat{a}_j^{(k)}/\hat{a}_k^{(k)'}\hat{a}_k^{(k)}, \quad j > k,$$

instead of (6.2.3).

To determine the column which leads to the least local fill-in, we can use the following theorem if $b_j^{(k)}$ denotes the jth column of a matrix B_k which is obtained from the last $n - k + 1$ columns of $A^{(k)}$ by replacing all the nonzero elements by unity.

(6.3.4)   *THEOREM*   If the $(t + k - 1)$th and kth columns of $A^{(k)}$ are interchanged and then the kth step of the RGS method is performed, then the maximum fill-in is given by the tth diagonal element of the matrix G_k, where

(6.3.5) $$G_k = (B_k' * B_k)\bar{B}_k' B_k,$$

where $*$ denotes Boolean multiplication and $\bar{B}_k$ is obtained from B_k by replacing all its zero elements by ones and vice versa.

Proof   If the $(t + k - 1)$th column of $A^{(k)}$ is made orthogonal to its $(j + k - 1)$th column, then the maximum number of additional nonzero elements created in the jth column is $\bar{b}_j^{(k)'}b_t^{(k)}$, where $\bar{b}_j^{(k)}$ is the jth column of $\bar{B}_k$. On the other hand, if $b_t^{(k)'} * b_j^{(k)} = 0$, then no nonzeros are created. Thus in both cases the total fill-in for the jth column is given by

$$(b_t^{(k)'} * b_j^{(k)})\bar{b}_j^{(k)'}b_t^{(k)}.$$

In view of the fact that $\bar{b}_t^{(k)'} b_t^{(k)} = 0$, the total fill-in for all the columns is given by

(6.3.6)
$$g_{tt}^{(k)} = \sum_{j=1}^{n-k+1} (b_t^{(k)'} * b_j^{(k)}) \bar{b}_j^{(k)'} b_t^{(k)}$$

$$= \sum_{j=1}^{n-k+1} (\tilde{e}_t' B_k' * B_k \tilde{e}_j) \tilde{e}_j' \bar{B}_k' B_k \tilde{e}_t,$$

where $\tilde{e}_j$ is the jth column of I_{n-k+1},

$$= \tilde{e}_t'(B_k' * B_k) \sum_{j=1}^{n-k+1} \tilde{e}_j \tilde{e}_j' \bar{B}_k' B_k \tilde{e}_t$$

$$= \tilde{e}_t'(B_k' * B_k) \bar{B}_k' B_k \tilde{e}_t,$$

since $\sum_{i=1}^{n-k+1} \tilde{e}_j \tilde{e}_j' = I_{n-k+1},$

$$= \tilde{e}_t' G_k \tilde{e}_t,$$

this completes the proof of the theorem.

We note that $b_p^{(k)'} b_q^{(k)} = 1$ does not necessarily imply that the inner product of the corresponding columns of $A^{(k)}$ is nonzero. Also the fill-in in $b_j^{(k)}$ may be less than $\bar{b}_j^{(k)'} b_t^{(k)}$, as in the actual RGS method cancellation may take place. This is the reason why $g_{tt}^{(k)}$ gives the maximum rather than the actual fill-in. It is our experience that such cases are rare, and when they occur, they represent only a small percentage of the overall computation; therefore the actual fill-in is generally very close to $g_{tt}^{(k)}$. In any case, it follows from the above theorem that in order to minimize the fill-in, we determine

(6.3.7)
$$g_{\hat{s}\hat{s}}^{(k)} = \min_t g_{tt}^{(k)} = \min_t [\tilde{e}_t' G_k \tilde{e}_t]$$

at the beginning of the kth step of the RGS method, then let $s = \hat{s} + k - 1$, and use (6.3.1), (6.3.2), (6.2.2), and (6.3.3) to compute $A^{(k+1)}$ from $A^{(k)}$. Note that only the diagonal elements of the product of the matrices $B_k' * B_k$ and $\bar{B}_k' B_k$ need to be computed to determine $\hat{s}$ at each stage k. Even this involves much work at each stage k. Therefore, we can arrange the columns of A a priori in ascending values of the diagonal elements of $(B_1' * B_1) \bar{B}_1' B_1$ before the RGS method is used. It is also possible to rearrange the columns of $A^{(k)}$ at periodic intervals rather than at each stage k by using Theorem 6.3.4. This generally leads to a further decrease in the fill-in. In order to determine the columns which yield a zero fill-in in other columns, the following corollaries to Theorem 6.3.4 are useful.

(6.3.8)   *COROLLARY*   If $b_t^{(k)'}b_t^{(k)} = b_j^{(k)'}b_t^{(k)}$ whenever $b_t^{(k)'} * b_j^{(k)} = 1$, then $g_{tt}^{(k)} = 0$.

Proof   Let V be an mth order column vector of all ones, then

$$0 = b_t^{(k)'}b_t^{(k)} - b_j^{(k)'}b_t^{(k)} = V'b_t^{(k)} - b_j^{(k)'}b_t^{(k)}$$
$$= (V' - b_j^{(k)'})b_t^{(k)}$$
$$= \bar{b}_j^{(k)'}b_t^{(k)}$$

and the corollary follows from (6.3.6).

(6.3.9)   *COROLLARY*   If $b_t^{(k)'}b_t^{(k)} = 1$, then $g_{tt}^{(k)} = 0$.

Proof   If $b_t^{(k)'}b_t^{(k)} = 1$, then $b_t^{(k)'} * b_j^{(k)} = 1$ implies that $b_t^{(k)'}b_j^{(k)} = 1$ and from Corollary 6.3.8, it follows that $g_{tt}^{(k)} = 0$.

Thus, in view of the above corollary, it is evident that all singleton columns should be orthonormalized prior to any of the other columns. The following theorem shows that a singleton column actually leads to a decrease in the total number of nonzero elements in those columns with which it interacts.

(6.3.10)   *THEOREM*   If $a_{pk}^{(k)} = 1$ but $a_{ik}^{(k)} = 0$ for all $i \neq p$, then $a_{pj}^{(k+1)} = 0$ for all $j > k$.

Proof   From (6.2.4), we have for $j > k$,

$$a_{pj}^{(k+1)} = a_{pj}^{(k)} - (a_{pk}^{(k)}a_{pj}^{(k)}/(a_{pk}^{(k)})^2)a_{pk}^{(k)}$$
$$= a_{pj}^{(k)} - a_{pj}^{(k)} = 0,$$

which completes the proof of the theorem.

Let V and V_k be column vectors of all ones of order m and $n - k + 1$, respectively, and e_i and $\tilde{e}_i$ denote the ith columns of I_n and I_{n-k+1}, respectively. Then, similar to (5.4.11), we define

(6.3.11) $r_i^{(k)} = e_i' B_k V_k$ and $c_j^{(k)} = V' B_k \tilde{e}_j.$

We can now describe an algorithm for the a priori arrangement of the columns of A to minimize the fill-in.

(6.3.12) *ALGORITHM*   Determine B_1 from A. Let R_1 denote a row vector consisting of first n natural numbers, namely, its jth element is the number j itself. Set $k = 1$.

1. Compute $c_j^{(k)}$ from (6.3.11). Find $c_t^{(k)} = \min_j c_j^{(k)}$, in the case of ties use the j with the lowest index. If $c_t^{(k)} > 1$, go to Step 2. Otherwise, set all $b_{pj}^{(k)} = 0$ where $b_{pt}^{(k)} = 1$ and replace the tth column of B_k by its first column and drop its first column from further consideration. Interchange the $(t + k - 1)$th and the kth elements of R_k. Set k to $k + 1$. If $k = n$, go to Step 3, otherwise go to the beginning of the current step.

2. Compute G_k according to (6.3.5) and rearrange the last $n - k + 1$ columns of R_k according to the ascending values of the diagonal elements of G_k and call the resulting vector R_{n+1}.

3. Rearrange the columns of A according to R_{n+1}, as the number in the jth position of R_{n+1} is the new position of the jth column of A.

Remarks   In Step 1 of the above algorithm, we determine all the singleton columns as well as those columns of A which may become singletons (according to Theorem 6.3.10), when singleton columns are made orthogonal to other columns. In Step 2, the remaining columns of A are arranged according to the amount of fill-in each would create if it was the column used at the end of Step 1. Note that in actual practice only a note of column permutation on A_k is kept in the course of the above algorithm and this information is used later when the RGS orthogonalization is performed.

The permutation matrices P and Q, such that $\hat{A} = PAQ$ is in a form which is desirable for the RGS method, can be found as in Chapter 3. If $\hat{A}$ is in any one of the forms: BDF, BTF, BNTF, SBBDF, DBBDF, BBTF or BBNTF, then the fill-in is limited to the shaded areas of these forms.

In the next section, we will discuss an orthogonal triangularization method which is due to Householder (1958). It utilizes elementary Hermitian matrices to transform A to an upper triangular form.

6.4. *The Householder Triangularization Method*

This method consists of n steps such that

(6.4.1) $$A^{(k+1)} = H_k A^{(k)}, \qquad k = 1, 2, \ldots, n$$

where

(6.4.2) $$H_k = I_m - \alpha_k^{-1} \hat{\eta}^{(k)} \hat{\eta}^{(k)'},$$

and the elements of the column vector $\hat{\eta}^{(k)}$ are given by

(6.4.3)
$$\hat{\eta}_i^{(k)} = 0, \quad i < k;$$
$$\hat{\eta}_k^{(k)} = a_{kk}^{(k)} \pm \beta_k, \quad \hat{\eta}_i^{(k)} = a_{ik}^{(k)}, \quad i > k$$

with

(6.4.4) $$\beta_k^2 = \sum_{i=k}^{m} (a_{ik}^{(k)})^2, \qquad \alpha_k = \beta_k^2 \pm \beta_k a_{kk}^{(k)}$$

and for stability β_k is chosen to have the same sign as $a_{kk}^{(k)}$. We start with $A^{(1)} \equiv A$ and at the end of n steps of the *Householder Triangularization* (HT) method, the first n rows of $A^{(n+1)}$ constitute an upper triangular matrix which we denote by $\bar{U}$ and its last $m - n$ rows are zero (we recall that $m \geqslant n$). Note that only $n - 1$ steps are needed in the HT method if $m = n$. Let

(6.4.5) $$H_n H_{n-1} \cdots H_1 = H$$

and the matrix constituting of the first n rows of H be denoted by $\hat{H}$, then from (6.4.1) and the fact that H is an orthogonal matrix, it follows that

$$A^{(n+1)} = HA \Rightarrow A = \hat{H}'\bar{U}$$

or

(6.4.6) $$A\bar{U}^{-1} = \hat{H}'.$$

Since $\bar{U}^{-1}$ is an upper triangular matrix and the columns of $\hat{H}'$ are orthonormal, therefore from (6.4.6), it follows that the HT method is an alternative way for orthonormalizing the columns of A. The high accuracy of the HT method makes it attractive for computation (Wilkinson 1965). Of course, the work involved is more than that for the RGS method. Matrix H is stored in the factored form (6.4.5); in fact only the nonzero elements of the $\hat{\eta}^{(k)}$s and the α_ks need to be stored (Tewarson, 1968a).

In order to discuss the fill-in for the HT method, we will need the following lemma:

(6.4.7) *LEMMA* If the kth step of the HT method is given by (6.4.1) through (6.4.4), then

    (i) the first $k - 1$ rows and columns of $A^{(k+1)}$ and $A^{(k)}$ are identical.

    (ii) $a_{kk}^{(k+1)} = \mp \beta_k$ and $a_{ik}^{(k+1)} = 0, i > k$.

Proof From (6.4.2) and (6.4.3), we see that the first $k - 1$ rows and columns of H_k are identical to the corresponding rows and columns of I_m; therefore, in view of (6.4.1), the first part of the lemma follows. Now, from (6.4.3) and (6.4.4), we have

$$\hat{\eta}^{(k)'} a_k^{(k)} = (a_{kk}^{(k)} \pm \beta_k) a_{kk}^{(k)} + \sum_{i=k+1}^{m} (a_{ik}^{(k)})^2$$

$$= \pm \beta_k a_{kk}^{(k)} + \beta_k^2 = \alpha_k,$$

and therefore, in view of (6.4.1) and (6.4.2), it follows that

$$a_k^{(k+1)} = a_k^{(k)} - \alpha_k^{-1}(\hat{\eta}^{(k)'} a_k^{(k)})\eta^{(k)}$$

$$= a_k^{(k)} - \hat{\eta}^{(k)} \Rightarrow a_{kk}^{(k+1)} = \mp \beta_k$$

and $a_{ik}^{(k+1)} = 0, i > k$. This completes the proof of the lemma.

From the above lemma it is clear that the only possible fill-in at the kth step of the HT is in the last $n - k + 1$ rows and $n - k$ columns of $A^{(k)}$. In order to minimize this fill-in, we first let B_k be the matrix obtained from the last $n - k + 1$ rows and columns of $A^{(k)}$ by replacing all the nonzero elements by unity and $b_j^{(k)} = B_k \tilde{e}_j$ and $\tilde{e}_i' B_k \tilde{e}_j = b_{ij}^{(k)}$ where $\tilde{e}_j$ is the jth column of I_{n-k+1}. Then we have the following:

(6.4.8)   *THEOREM*   If $b_{11}^{(k)} = 1$, then the maximum value of the fill-in at the kth step of the HT is given by the first diagonal element of the matrix G_k defined in (6.3.5).

   Proof   From (6.4.1), (6.4.2), (6.4.3), and (6.4.4), we have

$$a_q^{(k+1)} = a_q^{(k)} - \alpha_k^{-1}(\hat{\eta}^{(k)'} a_q^{(k)})\hat{\eta}^{(k)}, \qquad q > k.$$

But

$$\hat{\eta}^{(k)'} a_q^{(k)} = \sum_{i=k}^{m} a_{ik}^{(k)} a_{iq}^{(k)} \pm \beta_k a_{kq}^{(k)},$$

and since $b_{11}^{(k)} = 1$ ensures that $a_{kk}^{(k)} \neq 0$,

$$b_1^{(k)'} * b_j^{(k)} = 0 \Rightarrow \sum_{i=k}^{m} a_{ik}^{(k)} a_{iq}^{(k)} = 0 \Rightarrow \hat{\eta}^{(k)'} a_q^{(k)} = 0,$$

where $j = q - k + 1$. Consequently, in the qth column of $A^{(k)}$, there is no fill-in unless $b_1^{(k)'} * b_j^{(k)} = 1$, and in this case it cannot exceed $(b_1^{(k)'} * b_j^{(k)})\bar{b}_j^{(k)'} b_1^{(k)}$ where $\bar{b}_1^{(k)}$ is obtained from $b_1^{(k)}$ by changing all its unit elements to zero and vice versa. (Note that $b_1^{(k)'} * b_j^{(k)} = 1$ does not imply that $\hat{\eta}^{(k)} a_q^{(k)} \neq 0$, as cancellation of inner products may take place.) In view of the fact that $\bar{b}_t^{(k)'} b_t^{(k)} = 0$, the maximum fill-in in all the columns of $A^{(k)}$ is given by

$$g_{11}^{(k)} = \sum_{j=1}^{n-k+1} (b_1^{(k)'} * b_j^{(k)})\bar{b}_j^{(k)'} b_1^{(k)}$$

which is the same as (6.3.6) with $t = 1$. We conclude in a similar manner that

$$g_{11}^{(k)} = \tilde{e}_1' G_k \tilde{e}_1,$$

which completes the proof of the theorem.

Note that if $a_{kk}^{(k)} = 0$, then there may be a column v, for which $\hat{\eta}^{(k)'} a_v^{(k)} \neq 0$ but $a_k^{(k)'} a_v^{(k)} = 0$, and therefore unnecessary fill-in in column v can take place. This can be avoided by interchanging the kth and sth rows of $A^{(k)}$ prior to the kth step of the HT, where $a_{sk}^{(k)} \neq 0$. To take into account the row and column interchanges, we have the following.

(6.4.9) *THEOREM* If two columns of B_k are interchanged, then the corresponding diagonal elements of G_k have also to be interchanged. However, interchanging two rows of B_k has no effect on G_k.

Proof Let P_k and Q_k be the matrices obtained, respectively, from I_{n-k+1}, by interchanging its any two rows and any two columns. Now, if we replace B_k by $P_k B_k Q_k$, then the right-hand side of (6.3.5) is equal to

$$(Q_k' B_k' P_k' * P_k B_k Q_k)\, Q_k' \bar{B}_k' P_k' P_k B_k Q_k = Q_k' (B_k' * B_k) \bar{B}_k' B_k Q_k$$

$$= Q_k' G_k Q_k,$$

since

$$P_k' * P_k = Q_k Q_k' = Q_k * Q_k' = P_k' P_k = I_{n-k+1}.$$

This completes the proof.

From Theorems (6.4.8) and (6.4.9), it follows that in order to minimize the fill-in for the HT method, we determine, at the beginning of the kth stage, the index $\hat{s}$ according to (6.3.7) and another index $s \geqslant k$, such that $a_{s, \hat{s}+k-1}^{(k)} \neq 0$; then we interchange the kth and the $(\hat{s} + k - 1)$th columns and the kth and the sth rows of $A^{(k+1)}$. Evidently, the singleton columns of B_k do not lead to any fill-in and should be considered first. This is equivalent to a priori permutation of the rows and columns of A to get the largest upper triangular matrix in the top left-hand corner before starting the HT on the remaining matrix.

6.5. *The Fill-in for the RGS* versus *the HT Method*

In the previous section we observed that the HT can be performed on the given matrix A to obtain a matrix $\hat{H}'$ having orthonormal columns. The relation between A and $\hat{H}'$ is given by (6.4.6). If we recall that $\hat{H}'$ denotes the first n columns of $H_1'H_2' \cdots H_n'$ and store it in a factored form, then only the nonzero elements of the $\eta^{(k)}$s and α_ks are needed. Storing $\hat{H}'$ in a factored form generally involves no special problems, because $\hat{H}'$ is usually used later to multiply a vector (or a matrix). We will now show that $\hat{H}'$ in factored form is generally sparser than the matrix of orthonormal columns obtained by the RGS method.

In view of Theorems 6.3.4, 6.4.8, and 6.4.9, the maximum possible fill-in at the kth stage for both the RGS and the HT methods is given by the minimum diagonal elements of the corresponding G_k matrices. From (6.3.5) and the fact that the matrix B_k is $m \times (n - k + 1)$ for the RGS method but $(n - k + 1) \times (n - k + 1)$ for the HT, it is clear that at the kth stage of the RGS method generally more new nonzero elements are created than at the corresponding stage of the HT. Furthermore, in view of (6.4.3), only the last $n - k + 1$ nonzero elements of the kth column of $A^{(k)}$ are stored for the $\hat{\eta}^{(k)}$ vector in the HT, in contrast with the equivalent storage of the nonzero elements of the mth order vector, the kth column of $A^{(k)}$ in the RGS method. Therefore, it is evident that the factored form of $\hat{H}'$ is generally much sparser than the orthonormal columns obtained by the RGS method.

6.6. The Jacobi Method

Jacobi rotations, which are elementary orthogonal transformations (Wilkinson, 1965), can also be used to transform the given matrix to a matrix $A^{(n+1)}$, which is similar to the one obtained by the HT method. The Jacobi Method (JM) consists of $n - 1$ steps, each of which involves several minor steps. If $A^{(k)}$ denotes the matrix at the beginning of the kth major step, then the first $k - 1$ columns of $A^{(k)}$ are already in an upper triangular form. If $a_{ij}^{(k)}$ is the (i, j) element of $A^{(k)}$, then during the kth major step all $a_{ik}^{(k)} \neq 0$, $i > k$ are made zero. In each minor step, one $a_{ik}^{(k)} \neq$, $i > k$ is transformed to zero by a plane rotation. Thus the total number of minor steps in the kth major step is equal to the number of nonzero $a_{ik}^{(k)}$, $i > k$. Let us consider the first minor step of the kth major step. If $a_{pk}^{(k)}$ is the first nonzero element in column k which lies below the diagonal of $A^{(k)}$, then we define the orthogonal matrix R_{pk} as follows:

(6.6.1) $R_{pk} = I_n + (\tau - 1)(e_k e_k' + e_p e_p') + \omega(e_k e_p' - e_p e_k')$,

where

(6.6.2) $$\tau = a_{kk}^{(k)}/(a_{kk}^{(k)^2} + a_{pk}^{(k)^2})^{1/2}$$

and

$$\omega = a_{pk}^{(k)}/(a_{kk}^{(k)^2} + a_{pk}^{(k)^2})^{1/2}.$$

Thus R_{pk} is obtained from the identity matrix by replacing the (k, k), (k, p), (p, k), and (p, p) elements by τ, ω, $-\omega$, and τ, respectively. We will now show that all the rows of $A^{(k)}$ and $R_{kp}A^{(k)}$ are identical, except the kth and the pth rows which interact with each other and the (p, k) element of $R_{pk}A^{(k)}$ which is zero.

For $i \neq k$ or p, from (6.6.1) and the fact that $e_i'e_j = 0$, $i \neq j$, we have

(6.6.3) $e_i' R_{pk}A^{(k)} = e_i' A^{(k)}.$

On the other hand,

(6.6.4) $e_k' R_{pk}A^{(k)} = (e_k' + (\tau - 1)e_k' + \omega e_p')A^{(k)}$

$= \tau e_k' A^{(k)} + \omega e_p' A^{(k)}.$

Similarly,

(6.6.5) $e_p' R_{pk}A^{(k)} = (e_p' + (\tau - 1)e_p' - \omega e_k')A^{(k)}$

$= \tau e_p' A^{(k)} - \omega e_k' A^{(k)}.$

Thus the pth and the kth rows of $R_{pk}A^{(k)}$ are linear combinations of the corresponding rows of $A^{(k)}$. Finally, from (6.6.5) and (6.6.2) it follows that

(6.6.6) $e_p' R_{pk}A^{(k)}e_k = \tau a_{pk}^{(k)} - \omega a_{kk}^{(k)} = 0.$

From (6.6.4) and (6.6.5), it is evident that the fill-in takes place not only in the pth row but also in the kth row of $A^{(k)}$. Since the kth row of $R_{pk}A^{(k)}$ is used again in the next minor step to reduce some $a_{qk}^{(k)}$ to zero, where $q > p$, the nonzeros created in the first step in row k may also create nonzeros in row q. This will happen whenever $a_{kj}^{(k)} = 0$, $a_{pj}^{(k)} \neq 0$, and $a_{qj}^{(k)} = 0$. We call this *second order interaction* between rows p and q. Similarly, we have *third* and *higher order interactions* between the rows. Thus it is important to minimize the fill-in for row k in each minor step. As in the case of the HT and RGS methods, we consider the matrix B_k which is obtained from the last $n - k + 1$ rows and columns of A_k by replacing each nonzero element by unity. If the $(s + k - 1, t + k - 1)$ element of $A^{(k)}$ is moved to the (k, k) position at the beginning of the kth major step of the Jacobi method, then the fill-in can be determined from B_k, if cancellation in the computation is neglected. In any case, the actual fill-in, rather than the one determined from B_k, will be less if cancellation is taken into account. From (6.6.4) and (6.6.5), it is evident that the total fill-in for the kth step depends not only on row s of B_k but also on all the rows i of B_k for which $b_{it}^{(k)} = 1$, where $b_{it}^{(k)} = e_i'B_k e_t$. The fill-in will tend to be small if there are only a few rows for which $b_{it}^{(k)} = 1$, and these have few nonzero elements.

The total number of unit elements in all the rows for which $b_{it}^{(k)} = 1$ is given by (3.2.6), namely,

$$d_t^{(k)} = \sum_i r_i^{(k)}, \qquad \text{for all } i \text{ with } b_{it}^{(k)} = 1$$

$$= \sum_i b_{it}^{(k)} e_i' B_k V_k, \qquad \text{using (3.2.2)}$$

$$= \sum_i e_t' B_k' e_i e_i' B_k V_k$$

(6.6.7) $d_t^{(k)} = e_t' B_k' B_k V_k.$

Thus, in order to minimize the fill-in, we choose column t as follows:

(6.6.8) $d_t^{(k)} = \min_j d_j^{(k)},$

and then, to minimize the fill-in due to second and higher order interactions, we rearrange all the rows for which $b_{it}^{(k)} = 1$ in ascending values of $r_i^{(k)}$'s and

(6.6.9) $r_s^{(k)} = \min_i r_i^{(k)}, \qquad \text{for all } i \text{ with } \quad b_{it}^{(k)} = 1.$

It is possible to choose s and t by making use of the total fill-in for the kth row in the corresponding major step. Furthermore, the fill-in for the other rows can also be determined, but this requires too much work to be useful in practice.

The (s, j) element of B_k will be nonzero at the end of the kth step if

$$e_t' B_k' * B_k e_j = 1,$$

therefore the total number of new nonzero elements in row s is given by

(6.6.10)
$$\gamma_{st}^{(k)} = \sum_j e_t' B_k' * B_k e_j - r_s^{(k)}$$

$$= e_t'(B_k' * B_k)V_k - e_s'B_k V_k,$$

using (3.2.2). Thus, we can choose s and t as follows:

(6.6.11)
$$\gamma_{st}^{(k)} = \min_{i,j} \gamma_{ij}^{(k)}, \qquad \text{for all} \quad b_{ij}^{(k)} = 1.$$

6.7. *Bibliography and Comments*

The Householder triangularization and Jacobi methods with round-off error analyses are given in Wilkinson (1965). As a result of practical computational experiments, Rice (1966) has shown that the Revised Gram–Schmidt method yields better results than the ordinary Gram–Schmidt method. Round-off error analysis of the RGS method is given in Björck (1967). The use of orthonormalizing codes in numerical analysis is discussed by Davis (1962).

Eigenvalues and Eigenvectors

7.1. Introduction

There are two well-known direct methods for computing the eigenvalues and eigenvectors of symmetric matrices: the *Givens Method* (GM) and the *Householder Method* (HM). These are essentially equivalent to the Jacobi and the Householder methods for triangularization which were described in Chapter 6. In both methods, a series of orthogonal similarity transformations is used to modify the given matrix to a tridiagonal form, since the eigenvalues and eigenvectors of a tridiagonal matrix are easy to determine (Wilkinson, 1965; Fox 1965). In the next two sections, we shall give a brief description of these methods and delineate some techniques for minimizing the fill-in when the given matrix is transformed to a tridiagonal form (Tewarson, 1970a).

In the case of nonsymmetric matrices, a modification of the Gaussian elimination is used to transform the given matrix to a Hessenberg form, where all $a_{ij} = 0$, $i > j + 1$ (Wilkinson, 1965). The eigenvalues of a Hessenberg matrix are easy to find (Fox, 1965). In Section 7.4, a brief description of this method will be followed by techniques for minimizing the fill-in (Tewarson, 1970c).

If the given matrix A is symmetric, then in many cases, it is possible to permute it such that the top left-hand corner of the resulting matrix is in a tridiagonal form. This can be done, if we find a row having at most one off-diagonal element and move such a row (and the corresponding column) to become the first row (first column). Then we drop the first row and the first column from further consideration and repeat the above procedure for the remaining rows and columns. If at any stage no row having one off-diagonal element can be found, then we stop the process. Clearly, at this stage, the top left-hand corner of the transformed matrix is in a tridiagonal form. Now we have to transform only the square matrix in the bottom right-hand corner to a tridiagonal form by the GM or the HM. Therefore, without loss of generality, we shall denote this submatrix by A in the rest of this chapter.

7.2.   The Givens Method

This method reduces A to a tridiagonal matrix by Jacobi's rotations (Wilkinson, 1965). At the beginning of the kth major step, the matrix $A^{(k)}$ is tridiagonal as far as its first $k - 1$ rows and columns are concerned. The kth major step consists of at most $n - k - 1$ minor steps in which zeros are introduced successively in positions $k + 2$, $k + 3$, ..., n of the kth row and the kth column. Using a notation similar to that of Section 6.6, we define

$(7.2.1)$ $\qquad R_{pk} = I_n + (\tau - 1)(e_{k+1}e'_{k+1} + e_p e'_p)$

$\qquad\qquad\qquad + \omega(e_{k+1}e'_p - e_p e'_{k+1})$

where $a_{pk}^{(k)}$ is the first nonzero element after the $(k + 1)$th row of column k. Then the first minor step of the kth major step is

$(7.2.2)$ $\qquad\qquad\qquad A_1^{(k)} = R_{pk}A^{(k)}R'_{pk}.$

Now $e_i' R_{pk} = e_i'$ if $i \neq k + 1, p$; therefore, all the rows and columns, other than the $(k + 1)$th and the pth, of $A_1^{(k)}$ and $A^{(k)}$ are identical and

(7.2.3)
$$e_{k+1}' R_{pk} A^{(k)} = (e_{k+1}' + (\tau - 1)e_{k+1}' + \omega e_p')A^{(k)}$$
$$= \tau e_{k+1}' A^{(k)} + \omega e_p' A^{(k)},$$

and similarly,

(7.2.4)
$$e_p' R_{pk} A^{(k)} = \tau e_p' A^{(k)} - \omega e_{k+1}' A^{(k)}.$$

Thus the rows $k + 1$ and p of $R_{pk} A^{(k)}$ are linear combinations of the corresponding rows of $A^{(k)}$.

Similarly, we can see that in the matrix $R_{pk} A^{(k)} R_{pk}'$, columns $k + 1$ and p are linear combinations of the corresponding columns of $R_{pk} A^{(k)}$. Furthermore,

$$e_p' A_1^{(k)} e_k = e_p' R_{pk} A^{(k)} R_{pk}' e_k$$
$$= (\tau e_p' A^{(k)} - \omega e_{k+1}' A^{(k)})e_k$$
$$= \tau a_{pk}^{(k)} - \omega a_{k+1, k}^{(k)} = 0$$

if we take

$$\tau = a_{k+1, k}^{(k)} / [(a_{pk}^{(k)})^2 + (a_{k+1, k}^{(k)})^2]^{1/2}$$

and

$$\omega = a_{pk}^{(k)} / [(a_{pk}^{(k)})^2 + (a_{k+1, k}^{(k)})^2]^{1/2}.$$

Similarly, we can show that $e_k' A_1^{(k)} e_p = 0$. Thus the (p, k) and the (k, p) elements of $A_1^{(k)}$ are zero and by repeated use of (7.2.2), all the elements in positions $k + 2, k + 3, \ldots, n$ of the kth row and the kth column of $A^{(k)}$ are transformed to zero. The resulting matrix is denoted by $A^{(k+1)}$ and the kth major step is complete. We would like to determine an element $a_{s+k, t+k-1}^{(k)} \neq 0$, $s \neq t - 1$, such that if it is made the $(k + 1, k)$th element by symmetric row–column permutations, then the fill-in is minimized. This is a difficult combinatorial problem and its solution would require an excessive amount of computational effort to be useful in practice. Therefore, near optimal methods which do not require much work are generally preferred. The analysis given in Section 6.6 can also be applied here with minor modifications. The difference is that in Section 6.6, row k interacts with other rows, whereas here row $k + 1$ interacts with the other rows and then column $k + 1$ interacts with the other columns. Since $A^{(k)}$, $A_1^{(k)}$ and $A^{(k+1)}$ are all symmetric, minimizing the fill-in for row operations will generally minimize the column fill-in too. If B_k is the matrix obtained from the last $n - k$ rows and $n - k + 1$ columns of $A^{(k)}$ by replacing each nonzero element by unity, then either (6.6.8) and (6.6.9) or (6.6.11) can be used to determine s and t, such that $s \neq t - 1$ [as a diagonal element of $A^{(k)}$ cannot be made the $(k + 1, k)$ element by symmetric row–column permutations].

An interesting modification of the GM to band matrices is given by Schwarz (1968). In this method, a suitable sequence of rotations, analogous to the one given by (7.2.1), is applied to reduce a symmetric band matrix to a tridiagonal form. It preserves the band property of the given matrix throughout the whole process (see Fig. 7.2.1). The elimination of two symmetric elements within the band in general generates two identical nonzero elements in symmetric positions outside the band. These elements are eliminated by sequence of rotations

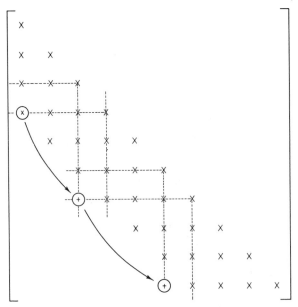

Fig. 7.2.1.   Rotations for a band matrix.

which shift them downwards λ rows and λ columns (where λ is the band width of the given matrix) and finally beyond the border of the matrix. In Fig. 7.2.1, for $n = 10$, $\lambda = 3$, the complete process for the elimination of element a_{41} is shown, where in view of the symmetry of A, only the elements on and below the main diagonal are considered. Let the rotation R_{pq} be defined as in (6.6.1), where τ and ω are chosen to transform a given element to zero. For example, in Fig. 7.2.1, first R_{34} is used to make a_{41} zero. This creates a nonzero in $(7, 3)$ position; R_{67} is used to make it zero. This in turn makes $(10, 6)$ element nonzero. $R_{9,10}$ is then used to make $(10, 6)$ element zero.

Similarly, $(3, 1)$ element is eliminated by R_{23} (which leaves $(4, 1)$ element unchanged). Two additional rotations will sweep the nonzero elements created in the sixth row and second column downwards and over the border of the matrix. A continuation of this process for other rows and columns will transform A to a tridiagonal form.

7.3. The Householder Method

This method reduces A to a triple diagonal matrix by elementary Hermitian orthogonal matrices (Wilkinson, 1965). At the beginning of the kth step, the matrix $A^{(k)}$ is tridiagonal in its first $k - 1$ rows and columns. In the kth step zeros are introduced in the kth row and the kth column without destroying the zeros already introduced in the previous steps. In other words, all elements in positions $k + 2, k + 3, \ldots, n$ of the kth row and the kth column are made zero. The method is similar to that described in Section 6.4. It consists of n steps such that $[A^{(1)} \equiv A$ and $A^{(n-1)}$ is triple diagonal$]$.

(7.3.1) $\qquad A^{(k+1)} = H_k A^{(k)} H_k, \qquad k = 1, 2, \ldots, n - 2,$

where

(7.3.2) $\qquad\qquad H_k = I_n - \alpha_k^{-1} \hat{\eta}^{(k)} \hat{\eta}^{(k)\prime},$

and the elements of the column vector $\hat{\eta}^{(k)}$ are given by

(7.3.3) $\qquad\qquad \hat{\eta}_i^{(k)} = 0, \quad i \leqslant k;$

$$\hat{\eta}_{k+1}^{(k)} = a_{k+1,k}^{(k)} \pm \beta_k, \hat{\eta}_i^{(k)} = a_{ik}^{(k)}; \quad i > k + 1$$

with

(7.3.4) $\beta_k{}^2 = \sum_{i=k+1}^{n} (a_{ik}^{(k)})^2, \qquad \alpha_k = \beta_k{}^2 \pm \beta_k a_{k+1,k}^{(k)}$

and for stability β_k is chosen to have the same sign as $a_{k+1,k}^{(k)}$.

Let B_k be the matrix obtained from the last $n - k$ rows and $n - k + 1$ columns of $A^{(k)}$ by replacing each nonzero element by unity. Then, analogous to Theorem 6.4.8, we have

(7.3.5)   *THEOREM*   If $b_{11}^{(k)} = 1$ and

(7.3.6) $\tilde{A}^{(k)} = H_k A^{(k)},$

where H_k is defined by (7.3.2), (7.3.3), and (7.3.4), then the maximum value for the fill-in is given by the $(1, 1)$ element of G_k defined in (6.3.5).

Proof   If we compare (7.3.6), (7.3.2), (7.3.3), and (7.3.4) with (6.4.1), (6.4.2), (6.4.3), and (6.4.4), respectively, then we see that the difference between the two sets of equations is that the kth row of $A^{(k)}$ is not used in the former. This is taken care of by constructing B_k (in this section) from the last $n - k$ rows rather than the last $n - k + 1$ rows as in Theorem 6.4.8. In view of this fact, the rest of the proof is the same as that of Theorem 6.4.8, and we will not repeat it here.

From the definition of G_k, it is evident that Theorem 6.4.9 applies also to the B_k defined in this section. Therefore, if we use (6.3.7) to select $\hat{s}$ and then choose $s \neq \hat{s} - 1$ such that $b_{s\hat{s}}^{(k)} = 1$ and move the $(s + k, \hat{s} + k - 1)$ element of $A^{(k)}$ to its $(k + 1, k)$ position before using (7.3.6), then the fill-in will be minimized. Since in (7.3.1) both $A^{(k+1)}$ and $A^{(k)}$ are symmetric, therefore the above choice for s and $\hat{s}$ which minimizes the fill-in when $A^{(k)}$ is multiplied by H_k, will also tend to minimize the fill-in when $H_k A^{(k)}$ is postmultiplied by H_k. It is possible to compute the actual fill-in in (7.3.1), but this generally involves too much computation to be advantageous in practice. In view of the above facts, we move the $(s + k, \hat{s} + k - 1)$ element of $A^{(k)}$ to the $(k + 1, k)$ position by symmetric row–column permutations prior to the kth step of the HM. This will keep the fill-in small.

A simpler way of finding $\hat{s}$ is to select a column of B_k which interacts with few other columns; in other words such a column has a nonzero Boolean inner product with a minimum number of other columns, namely,

$$(7.3.7) \qquad e_{\hat{s}}{}'(B_k{}' * B_k)V = \min_j e_j{}'(B_k{}' * B_k)V.$$

Of course, this would generally give a less accurate estimate for fill-in than the use of (6.3.7).

So far in this chapter, we have discussed the Givens method and the Householder method for symmetric sparse matrices. In the next section, we will discuss how an unsymmetric matrix can be transformed to an upper Hessenberg form by elementary similarity transformations.

7.4. *Reduction to the Hessenberg Form*

Let $A^{(k)}$ be the matrix with its first $k - 1$ columns in Hessenberg form, namely, $a_{ij}^{(k)} = 0$ for all $i > j + 1$ and $j < k$. Then at the kth stage elementary similarity transformations (which are very much like the matrices L_k of Section 2.2) are used to make the elements in positions $k + 2, k + 3, \ldots, n$ of the kth column zero. This is done for $k = 1, 2, \ldots, n - 2$ with the result that $A^{(n-1)}$ is in Hessenberg form. Thus we have

$$(7.4.1) \qquad A^{(k+1)} = L_{k+1} A^{(k)} L_{k+1}^{-1}, \qquad k = 1, 2, \ldots, n - 2,$$

where

$$(7.4.2) \qquad L_{k+1} = I_n + \eta^{(k+1)} e_{k+1}'$$

and the elements of the column vector $\eta^{(k)}$ are given by

$$(7.4.3) \qquad \begin{aligned} \eta_i^{(k+1)} &= 0, \quad i \leqslant k + 1; \\ \eta_i^{(k+1)} &= -a_{ik}^{(k)}/a_{k+1,k}^{(k)}, \quad i > k + 1. \end{aligned}$$

From (7.4.2) and (7.4.3), it follows that

$$(7.4.4) \qquad L_{k+1}^{-1} = I_n - \eta^{(k+1)} e_{k+1}'.$$

The above equations imply that $A^{(k+1)}$ is obtained from $A^{(k)}$ by first adding appropriate multiples of the $(k+1)$th row to all rows with $a_{ik}^{(k)} \neq 0$, $i > k+1$, and then adding multiples of all columns j of the resulting matrix for which $a_{jk}^{(k)} \neq 0$, $j > k+1$, to the $(k+1)$th column. Thus fill-in may take place in the whole $(k+1)$th column and in the last $n-k-1$ rows and columns of $A^{(k)}$. In the last $n-k-1$ components of the $(k+1)$th column, the fill-in can take place twice: once due to the premultiplication by L_{k+1}, and once due to the postmultiplication by L_{k+1}^{-1}. Of course, in the first $k+1$ components of the $(k+1)$th column, fill-in takes place only due to L_{k+1}^{-1}. On the other hand, for the last $n-k-1$ rows and columns, fill-in can take place only due to L_{k+1}. Let B_k denote the matrix obtained from the last $n-k$ rows and $n-k+1$ columns of $A^{(k)}$ by replacing each nonzero element by unity. Then we have

(7.4.5)   *THEOREM*   If the $(s+k, t+k-1)$ element of $A^{(k)}$ is moved to the $(k+1, k)$ position by symmetric row–column permutations, where $s \neq t-1$, and the resulting matrix is premultiplied by L_{k+1}, then the maximum fill-in is given by the (s, t) element of G_k defined by (2.5.6).

Proof   The restriction $s \neq t-1$ is needed because a diagonal element of $A^{(k)}$ cannot be moved to the $(k+1, k)$ position by symmetric row–column permutations. If we note the fact that the (i, j) element of B_k corresponds to the $(i+k, j+k-1)$ element of $A^{(k)}$, then the proof is the same as that of Theorem 2.5.5, and there is no need to repeat it here.

Since L_{k+1}^{-1} only changes the $(k+1)$th column of $A^{(k)}$, and if we neglect this fill-in, then Theorem 7.4.5 can be used to select a pivot to minimize the fill-in at each stage. It is possible to compute the fill-in for the $(k+1)$th column due to L_{k+1}^{-1}, but this involves too much work to be useful in practice.

We will now describe how the fill-in for the $(k+1)$th column can be minimized, if $A^{(k)}$ is postmultiplied only by L_{k+1}^{-1}. Let N_k denote the zero–one matrix obtained from the last $n-k+1$ columns of $A^{(k)}$ by replacing each nonzero element by unity. Let $\tilde{B}_k$ denote the matrix consisting of the only last $n - k + 1$ rows of N_k. Furthermore, let $I^{(q)}$ denote the matrix obtained from the identity matrix of order $n - k + 1$ by replacing the qth diagonal element by zero. Then we have

(7.4.6) *THEOREM* If the $(p + k - 1, q + k - 1)$ element of $A^{(k)}$ is moved to the $(k + 1, k)$ position by symmetric row–column permutations and the resulting matrix is multiplied by L_{k+1}^{-1}, then the maximum fill-in in the $(k + 1)$th column is given by

(7.4.7)
$$\gamma_{pq}^{(k)} = e_p' N_k (N_k * I^{(q)} \tilde{B}_k) e_q.$$

Proof From (7.4.3), (7.4.4), the definitions of N_k and $\tilde{B}_k$, and the fact that in $A^{(k)}$ columns $q + k - 1$ and $p + k - 1$ become, respectively, columns k and $k + 1$, we have

$$\gamma_{pq}^{(k)} = (\overline{N}_k e_p)' \left(\sum_{i \neq q}^* \tilde{b}_{iq}^{(k)} N_k e_i \right),$$

where $\tilde{b}_{iq}^{(k)}$ is the (i, q) element of $\tilde{B}_k$ and $\sum^*$ denotes the Boolean sum of the columns. Now,

$$\gamma_{pq}^{(k)} = e_p' \overline{N}_k' \left(N_k * \sum_{i \neq q} e_i e_i' \tilde{B}_k e_q \right)$$
$$= e_p' \overline{N}_k' (N_k * I^{(q)} \tilde{B}_k) e_q,$$

which completes the proof of the theorem.

In view of Theorems 7.4.5 and 7.4.6, and the fact that the (i, j) element of B_k is the same as the $(i + 1, j)$ element of $\tilde{B}_k$, it is evident that we can choose the pivot $a_{s+k, t+k-1}^{(k)}$ as follows:

(7.4.8)
$$g_{st}^{(k)} + \gamma_{s+1, t}^{(k)} = \min_{i,j} (g_{ij}^{(k)} + \gamma_{i+1, j}^{(k)}), \qquad i \neq j + 1,$$

where $g_{ij}^{(k)}$ is the (i, j) element of $G^{(k)}$. This should, in general, minimize the fill-in.

A simpler, though less accurate way to choose the pivot, is to note the fact that the fill-in depends on the total number of nonzero elements in row s and column t. Therefore, we choose s and t as follows: Let B_k be defined as for Theorem 7.4.5, and V_k as in (3.2.2) but V_k' be an $n - k$

element row vector of all ones. Then, in view of (3.2.2) and (3.2.3),

$$(7.4.9) \qquad\qquad \hat{g}_{st}^{(k)} = \min_{i,j} \hat{g}_{ij}^{(k)}, \qquad i \neq j - 1,$$

can be used to determine s and t.

7.5. *Eigenvectors*

The eigenvector x corresponding to a known eigenvalue λ can be easily obtained, because $Ax = \lambda x$ implies that

$$(7.5.1) \qquad\qquad (A - \lambda I)x = 0.$$

Note that $A - \lambda I$ is singular, since $x \neq 0$, and therefore we could omit any equation from (7.5.1) and solve the rest as a set of nonhomogeneous equations in the $(n - 1)$ ratios of the components of x. Round-off and other computational considerations for this are mentioned in Fox (1965). In the solution of the nonhomogeneous set of equations, the various techniques for minimizing the fill-in and/or computational effort given in the previous chapters can be used.

7.6. *Bibliography and Comments*

The various direct methods for computing the eigenvalues and eigenvectors of full matrices and error analyses are given in Fox (1965) and Wilkinson (1965). The Jacobi rotations for matrices in band form are described in Rutishauser (1963) and Schwarz (1968). The fill-in for Givens' and Householder's methods is discussed in Tewarson (1970a). Also in Tewarson (1970c) are given some techniques for minimizing the fill-in when reducing a given matrix to the Hessenberg form.

8

Change of Basis and Miscellaneous Topics

8.1. Introduction

In some practical applications, modifications have to be made to the given matrix A after its inverse (PFI or EFI) has been determined. This occurs, for example, in linear programming, where, at each step of the simplex method, one column in the "basis" is replaced by a "nonbasic" column, and the inverse of the basis is updated to become the inverse of the modified basis (Orchard-Hays, 1968). Another example, from electrical networks, is known as Kron's method of tearing (Kron, 1963), where the change in A is given by a matrix of small rank. In this chapter, we will discuss several methods for incorporating the result of modifications to A in its inverse (Dantzig, 1963b; Bartels and Golub, 1969; Brayton *et al.*, 1969; Forrest and Tomlin, 1972). In Section 8.2, we describe the various methods for modifying the EFI and PFI if one column of the given matrix is changed (the changes in the rows of A can be incorporated by considering the changes in the corresponding columns of A'). Kron's method of tearing for sparse matrices is described in Section 8.3. A factored form of inverse of A is given in Section 8.4. It results when the matrix U is inverted in a different manner than in Section 2.2.

8.2.   The Result of Changes in a Column of A on A⁻¹

Suppose the EFI, PFI, or some other form of inverse of A is available. The EFI and the PFI are given by (2.4.1) and (5.2.4), respectively. Let $\hat{A}$ denote the matrix obtained from A by changing the qth column of A to a new column, say $\hat{a}_q$. We shall now describe several ways of how $\hat{A}^{-1}$ can be constructed from A^{-1}.

FIRST METHOD

If A^{-1} denotes a form of the inverse of A, then each column of $A^{-1}\hat{A}$ is identical to the corresponding column of I_n, except the qth column, namely,

$$(8.2.1) \qquad A^{-1}\hat{A} = I_n + (A^{-1}\hat{a}_q - e_q)e_q'$$
$$= I_n + (\hat{a}_q^{(n+1)} - e_q)e_q',$$
$$\text{where} \quad \hat{a}_q^{(n+1)} = A^{-1}\hat{a}_q.$$

Therefore,

$$(8.2.2) \qquad \hat{A}^{-1} = [I_n + (\hat{a}_q^{(n+1)} - e_q)e_q']^{-1}A^{-1}$$
$$= \hat{T}_q A^{-1},$$

where, in view of Section 5.2,

$$(8.2.3) \qquad \hat{T}_q = I_n + (\hat{\zeta}^{(q)} - e_q)e_q'$$

with

$$(8.2.4) \qquad \hat{\zeta}_i^{(q)} = -\hat{a}_{iq}^{(n+1)}/\hat{a}_{qq}^{(n+1)}, \quad i \neq q \quad \text{and} \quad \hat{\zeta}_q^{(q)} = 1/\hat{a}_{qq}^{(n+1)}.$$

Thus $\hat{A}^{-1}$ has one additional factor $\hat{T}_q$ than the factored form of A^{-1}. We recall that only the nonzero elements of $\hat{\zeta}^{(q)}$ need be stored for evaluating $\hat{T}_q$.

Additional columns of A can be changed in a similar manner. Of course, each such change adds another factor (similar to $\hat{T}_q$) to the inverse. If only a few columns need to be modified, then the present method is reasonable to use. On the other hand, if several columns of A are changed, as in linear programming, then we would like to get rid of those factors of A^{-1} which correspond to the original columns of A that have been changed. Each of these columns can be thought of as being removed from the "basis" and a new column (which is the changed column) inserted in its place. In the next two methods, which work only for the EFI, the factor corresponding to the column removed from the basis is deleted.

SECOND METHOD

We will show that changing a_q to $\hat{a}_q$ results in the replacement of U_q in (2.4.1) by a matrix $\hat{T}_q$ defined by (8.2.3), with $\hat{\zeta}^{(q)}$ given by

(8.2.5) $\qquad \zeta_i^{(q)} = -\hat{a}_{iq}^{(t)}/\hat{a}_{qq}^{(t)}, \qquad i \neq q \qquad$ and $\qquad \hat{\zeta}_q^{(q)} = 1/\hat{a}_{qq}^{(t)}$

where

(8.2.6) $\qquad \hat{a}_q^{(t)} = U_{q+1} \cdots U_n L_n \cdots L_1 \hat{a}_q.$

It is evident that, except for its qth column, the matrix $L_n \cdots L_1 \hat{A}$ is identical to the upper triangular matrix $A^{(n+1)}$ of (2.2.5). Let the qth column of $L_n \cdots L_1 \hat{A}$ be denoted by $\hat{a}_q^{(n+1)}$. The factored form of the inverse of the matrix $L_n \cdots L_1 \hat{A}$ is obtained, as in Section 2.2, by taking the pivots on the leading diagonal. Since $L_n \cdots L_1 A$ and $L_n \cdots L_1 \hat{A}$ differ only in the qth column, and, in view of (2.2.9), (2.2.10), and (2.2.11), $U_{q+1} \cdots U_n L_n \cdots L_1 \hat{A}$ and $U_{q+1} \cdots U_n L_n \cdots L_1 A$ are identical, except for the qth column. The qth column of the former matrix is given by (8.2.6). It is evident from (8.2.3), (8.2.5), and (8.2.6) that $\hat{T}_q$ will transform the qth column of $U_{q+1} \cdots U_n L_n \cdots L_1 \hat{A}$ to e_q and not affect the other columns; in other words, matrices $\hat{T}_q U_{q+1} \cdots U_n L_n \cdots L_1 \hat{A}$

and $U_q U_{q+1} \cdots U_n L_n \cdots L_1 A$ are identical. Therefore,

$$I_n = U_2 \cdots U_n L_n \cdots L_1 A$$

$$\equiv U_2 \cdots U_{q-1} \hat{T}_q U_{q+1} \cdots U_n L_n \cdots L_1 \hat{A}$$

and

(8.2.7) $\hat{A}^{-1} = U_2 \cdots U_{q-1} \hat{T}_q U_{q+1} \cdots U_n L_n \cdots L_1.$

To change another column q_1 after column q has been altered, we have two cases:

1. If $q_1 \leqslant q$, then $\hat{T}_{q_1}$ replaces U_{q_1} and

$$\hat{a}_{q_1}^{(t)} = U_{q_1+1} \cdots U_{q-1} \hat{T}_q U_{q+1} \cdots U_n L_n \cdots L_1 \hat{a}_{q_1}.$$

2. If $q_1 > q$, then $\hat{T}_{q_1}$ is inserted immediately after $\hat{T}_q$, where

$$\hat{a}_{q_1}^{(t)} = \hat{T}_q U_{q+1} \cdots U_n L_n \cdots L_1 \hat{a}_{q_1},$$

and U_{q_1} is not replaced.

From the above two cases, it is clear that the columns should be replaced in descending order of their indices if possible (Brayton *et al.*, 1969).

THIRD METHOD

This method is especially suited to linear programming codes (Brayton *et al.*, 1969; Forrest and Tomlin, 1972). As before, let a_q, the qth column of A, be replaced by $\hat{a}_q$ and $\hat{A}$ denote the modified matrix. If $\hat{A}^{(n+1)} = L_n \cdots L_1 \hat{A}$ and $U = L_n \cdots L_1 A$, then only the qth columns of $\hat{A}^{(n+1)}$ and U are different. Now, elementary matrices $\hat{U}_q$ and $\hat{T}_q$ are determined, such that the last $n - q$ elements of the qth row of $\hat{U}_q \hat{A}^{(n+1)}$ are zero and e_q is the qth column of $U^{(q)} = \hat{T}_q \hat{U}_q \hat{A}^{(n+1)}$. Evidently, $U^{(q)}$ is easy to invert, because it is obtained from U by replacing its qth row and column by e_q' and e_q, respectively. Thus, in view of (2.2.9), (2.2.10), and (2.2.11), $U^{(q)-1}$ is obtained from $U_2 \cdots U_n$ by deleting U_q and setting each $\zeta_q^{(k)} = 0$, $k > q$, in (2.2.11). Clearly,

(8.2.8) $\hat{A}^{-1} = U^{(q)-1} \hat{T}_q \hat{U}_q L_n \cdots L_1.$

We have to determine $\hat{U}_q$ and $\hat{T}_q$ to make use of the above formula for $\hat{A}^{-1}$. This can be done as follows. If

$$(8.2.9) \qquad \hat{U}_q = I_n + e_q \hat{\xi}^{(q)},$$

where

$$e_q' + \hat{\xi}^{(q)} = e_q' U_{q+1} \cdots U_n$$

then, in view of the fact that $\hat{A}^{(n+1)} e_j = U e_j, j \neq q$, we have

$$e_q' \hat{U}_q \hat{A}^{(n+1)} e_j = (e_q' + \hat{\xi}^{(q)}) U e_j$$
$$= e_q' U_{q+1} \cdots U_n U e_j$$
$$= e_q' e_j = 0, \qquad j \neq q,$$

since $U_{q+1}, \ldots, U_n$ transform the last $n - q$ columns of U to the corresponding columns of I_n (see Section 2.2). Furthermore, since $e_j' \hat{U}_q = e_j', j \neq q$, $\hat{U}_q$ given by (8.2.9) is the desired matrix which reduces all nondiagonal elements of the qth row of $\hat{A}^{(n+1)}$ to zero and leaves the other rows unchanged.

Let

$$(8.2.10) \qquad \hat{a}_q^{(t)} = \hat{U}_q L_n \cdots L_1 \hat{a}_q,$$

then, since $e_q' \hat{U}_q \hat{A}^{(n+1)} = \hat{a}_{qq}^{(t)} e_q'$, the Gauss–Jordan elimination performed on the qth column of $\hat{U}_q \hat{A}^{(n+1)}$ will not affect any of the other columns, and matrix $\hat{T}_q$ is the same as (8.2.3) with $\hat{\xi}^{(q)}$ defined by (8.2.5) and $\hat{a}_q^{(t)}$ as in (8.2.10).

In order to change a column of $\hat{A}$, say $\hat{a}_{q_1}$ to $\hat{\hat{a}}_{q_1}$, it can be easily checked that

$$(8.2.11) \qquad \hat{\hat{A}}^{-1} = U^{(q,q_1)-1} \hat{T}_{q_1} \hat{U}_{q_1} \hat{T}_q \hat{U}_q L_n \cdots L_1,$$

where $U^{(q,q_1)}$ is obtained from $U^{(q)}$ by replacing its qth row and column by e_{q_1}' and e_{q_1}, respectively, and $\hat{T}_{q_1}$ and $\hat{U}_{q_1}$ are obtained from $\hat{T}_q \hat{U}_q L_n \cdots L_1 \hat{A}$, in the same manner as $\hat{T}_q$ and $\hat{U}_q$ resulted from $L_n \cdots L_1 \hat{A}$. It is not necessary to transform the qth column of $\hat{U}_q \hat{A}^{(n+1)}$ to the unit vector e_q, if it is recorded that when inverting $\hat{U}_q \hat{A}^{(n+1)}$ its qth column should be first transformed to a unit vector (Forrest and Tomlin, 1972). In this case, $\hat{T}_q$ and $\hat{T}_{q_1}$ are absent from (8.2.11); however, when inverting the matrix $U^{(q,q_1)}$ columns q and q_1 (one with the smaller index first) should be reduced to unit vectors prior to the rest of the matrix.

8.3. Kron's Method of Tearing

Let K, E, and C denote, respectively, matrices of order $n \times r, r \times r$, and $r \times n$, and

$$\hat{A} = A + KEC,$$

then it can be easily verified by direct multiplication that

(8.3.1) $\hat{A}^{-1} = [I_n - A^{-1}KE(I_r + CA^{-1}KE)^{-1}C]A^{-1}.$

If A^{-1} is available in a factored form, then $\hat{A}^{-1}$ can be computed as follows:

   1. Compute the $n \times r$ matrix $Y = A^{-1}KE$.

   2. Solve $(I_r + CY)'Z' = Y'$, for the $r \times n$ matrix Z'.

   3. Compute $\hat{A}^{-1} = (I_n - ZC)A^{-1}$.

To store $\hat{A}^{-1}$, we only have to store Z, C, and A^{-1}. We will now show that the first method of the previous section is a special case of Kron's method. If $K = \hat{a}_q - a_q$, $E = 1$ and $C = e_q'$, then the matrix which premultiplies A^{-1} in (8.3.1) is equal to

$$I_n - A^{-1}(\hat{a}_q - a_q)[1 + e_q'A^{-1}(\hat{a}_q - a_q)]^{-1}e_q'$$
$$= I_n - (\hat{a}_q^{(n+1)} - e_q)[1 + e_q'(\hat{a}_q^{(n+1)} - e_q)]^{-1}e_q'$$
$$= I_n - (\hat{a}_q^{(n+1)} - e_q)[\hat{a}_{qq}^{(n+1)}]^{-1}e_q'$$
$$= \hat{T}_q,$$

in view of (8.2.3) and (8.2.4). Thus, (8.3.1) becomes the same as (8.2.2).

   In the next section, we will show how the matrices of the type (8.2.9) can be used to express A^{-1} in a factored form (Zollenkopf, 1971).

8.4. Bifactorization

The matrix U obtained at the end of the forward course of the Gaussian elimination in Section 2.2 can be transformed to I_n by using elementary column operations, such that

$$(8.4.1) \qquad U\hat{U}_1\hat{U}_2\cdots\hat{U}_{n-1} = I_n$$

where, for $k = 1,\ldots,n-1$, $\hat{U}_k$ transforms the kth row of $U\hat{U}_1\cdots\hat{U}_{k-1}$ to e_k' by subtracting multiples of the kth column of $U\hat{U}_1\cdots\hat{U}_{k-1}$, which is equal to e_k, from the succeeding columns. Evidently, all the other rows of $U\hat{U}_1\cdots\hat{U}_{k-1}$ remain unchanged and the last $n-k$ rows are the same as those of U. Therefore, $\hat{U}_k$ is given by

$$(8.4.2) \qquad \hat{U}_k = I_n + e_k\hat{\xi}^{(k)}$$

where

$$(8.4.3) \qquad \hat{\xi}_j^{(k)} = 0, \quad j \leqslant k, \qquad \text{and} \qquad \hat{\xi}_j^{(k)} = -u_{kj}, \quad j > k.$$

From (8.4.1), we have $(\hat{U}_1\cdots\hat{U}_{n-1})U = I_n$, and therefore, in view of (2.2.6) and (2.2.7), it follows that

$$\hat{U}_1\cdots\hat{U}_{n-1}L_n\cdots L_1 A = I_n,$$

which gives

$$(8.4.4) \qquad A^{-1} = \hat{U}_1\cdots\hat{U}_{n-1}L_n\cdots L_1.$$

Since $A^{(k+1)}$ is defined by (2.2.2), (2.2.3), and (2.2.4), and U by (2.2.6) and (2.2.7), the first k rows of both the matrices are therefore identical. From this fact and (8.4.2) and (8.4.3), it follows that $\hat{U}_k$ can be computed as soon as $A^{(k+1)}$ is known. In other words, the L_ks and $\hat{U}_k$s can be evaluated in the following order

$$L_1, \hat{U}_1, L_2, \hat{U}_2, \ldots, L_{n-1}, \hat{U}_{n-1}, L_n$$

and there is no back substitution of Section 2.2.

8.5.   *Bibliography and Comments*

The first method described in Section 8.2 is well known in linear programming, if the "basis" inverse is stored in product form (Dantzig, 1963a). Packing schemes for storing the product form of inverse (PFI) are given in Smith (1969) and de Buchet (1971). The elimination form of inverse (EFI), which is needed in the second and third methods, was first advocated by Markowitz (1957) and later by Dantzig (1963b) for specially structured "staircase" matrices. Dantzig *et al.* (1969) demonstrated the superiority of the EFI over the PFI for general linear programming matrices in terms of speed, accuracy, and sparsity of the resulting matrix. Updating the EFI has been investigated by Dantzig (1963b) for his "staircase" algorithm in such a way that the special structure of factors is preserved. Bartels and Golub (1969) have proposed a triangular updating scheme having certain desirable properties; however, Forrest and Tomlin (1972) observe that there are some practical difficulties in implementing the scheme for large-scale sparse linear programs. It turns out that the third method of Section 8.2, which is due to Brayton *et al.* (1969) and implemented and developed for linear programs by Tomlin (1970) and Forrest and Tomlin (1972), was found to be the most suitable for large-scale sparse linear programs. This method is presently being used to solve real practical linear programming problems (Forrest and Tomlin, 1972).

If the first and the second methods of Section 8.2 are compared, then, in view of (8.2.1)–(8.2.4) and (8.2.5)–(8.2.7), we can conclude that $\hat{T}_q$ in the first method not only requires more work, but also tends to be denser than $\hat{T}_q$ in the second method. Furthermore, in the second method, U_q is replaced by $\hat{T}_q$, in contrast with the first method where $\hat{T}_q$ becomes an additional factor of A^{-1}. Thus the second method is generally better than the first one.

When solving sparse nonlinear systems by quasi-Newton type methods, the determination of single rank corrections for a sparse matrix such that the resulting matrix is also sparse, but is a better approximation for the Jacobian is discussed in Schubert (1970) and Broyden (1971).

Kron's method (Kron, 1963) is also described in Roth (1959) and Spillers (1968).

References

Not all references are cited in the text.

Akyuz, F. A., and Utku, S. (1968). An automatic relabeling scheme for bandwidth minimization of stiffness matrices. *AIAA J.* **6**, 728–730.

Allwood, R. J. (1971). Matrix methods of structural analysis. *In* "Large Sparse Sets of Linear Equations" (J. K. Reid, ed.), pp. 17–24. Academic Press, New York

Alway, G. G., and Martin, D. W. (1965). An algorithm for reducing the bandwidth of a matrix of symmetrical configuration. *Comput. J.* **8**, 264–272.

Arany, I., Smyth, W. F., and Szoda, L. (1971). An improved method for reducing the bandwidth of sparse, symmetric matrices. *IFIP Conf., Ljubljana, Yugoslavia.*

Ashkenazi, V. (1971). Geodetic normal equations. *In* "Large Sparse Sets of Linear Equations" (J. K. Reid, ed.), pp. 57–74. Academic Press, New York.

Baker, J. M. (1962). A note on multiplying Boolean matrices. *Comm. ACM* **5**, 102.

Bartels, R. H., and Golub, G. H. (1969). The simplex method of linear programming using LU decomposition. *Comm. ACM* **12**, 266–268.

Baty, J. P., and Stewart, K. L. (1971). Organization of network equations using disection theory. *In* "Large Sparse Sets of Linear Equations" (J. K. Reid, ed.), pp. 169–190. Academic Press, New York.

Baumann, R. (1965). Some new aspects of load flow calculation. *IEEE Trans. Power Apparatus and Systems* **85**, 1164–1176.

141

Baumann, R. (1971). Sparseness in power systems equations. *In* "Large Sparse Sets of Linear Equations" (J. K. Reid, ed.), pp. 105–126. Academic Press, New York.

Bauer, F. L. (1963). Optimally scaled matrices. *Numer. Math.* **5**, 73–87.

Beale, E. M. L. (1971). Sparseness in linear programming. *In* "Large Sparse Sets of Linear Equations" (J. K. Reid, ed.), pp. 1–16. Academic Press, New York.

Bellman, R., Cooke, K. L., and Lockett, J. A. (1970). "Algorithms, Graphs and Computers." Academic Press, New York.

Benders, J. F. (1962). Partitioning procedures for solving mixed-variable programming problems. *Numer. Math.* **4**, 238–252.

Berry, R. D. (1971). An optimal ordering of electronic circuit equations for a sparse matrix solution. *IEEE Trans. Circuit Theory* **CT-18**, 40–50.

Bertelé, U., and Brioschi, F. (1971). On the theory of the elimination process. *J. Math. Anal. Appl.* **35**, 48–57.

Björck, A. (1967). Solving linear least squares problems by Gram–Schmidt orthogonalization. *Nordisk Tidskr. Informations-Behandling* (*BIT*) **7**, 1–21.

Branin, F. H., Jr. (1959). The relation between Kron's method and the classical methods of network analysis. WESCON Convention Record (Part 2), 1–29.

Branin, F. H., Jr. (1967). Computer methods of network analysis. *Proc. IEEE* **55**, 1787–1801.

Brayton, R. K., Gustavson, F. G., and Willoughby, R. A. (1969). Some results on sparse matrices. Rep. No. RC 2332, IBM, Yorktown Heights, New York. (A shorter version of this appeared in 1970 in *Math. Comput.* **24**, 937–954.)

Bree, D., Jr. (1965). Some remarks on the application of graph theory to the solution of sparse systems of linear equations. Ph.D. thesis. Math. Dept., Princeton Univ., Princeton, New Jersey.

Broyden, C. G. (1971). The convergence of an algorithm for solving sparse non-linear systems. *Math. Comput.* **25**, 285–294.

Buchet, J. de (1971). How to take into account the low density of matrices to design a mathematical programming package. *In* "Large Sparse Sets of Linear Equation" (J. K. Reid, ed.), pp. 211–218. Academic Press, New York.

Bunch, J. R. (1969). On direct methods for solving symmetric systems of linear equations. Ph.D. thesis, Univ. of California, Berkeley, California.

Busacker, R. G., and Saaty, T. L. (1965). "Finite Graphs and Networks." McGraw-Hill, New York.

Cantin, G. (1971). An equation solver of very large capacity. *Internt. J. Numer. Methods Eng.* **3**, 379–388.

Carpentier, J. (1963). Ordered eliminations. *Proc. Power Systems Comput. Conf., London.*

Carpentier, J. (1965). Éliminations ordonnées—un processus diminuant le volume des calculs dans la résolutions des systèmes linéaires a matrice creuse. "Troisième Congr. de Calcul et de Traitement de l'Information AFCALTI," pp. 63–71. Dunod, Paris.

Carré, B. A. (1971). An elimination method for minimal-cost network flow problems. *In* "Large Sparse Sets of Linear Equations" (J. K. Reid, ed., pp. 191–210. Academic Press, New York.

Carré, B. A. (1966). The partitioning of network equations for block iterations. *Comput. J.* **9**, 84–97.

Chang, A. (1969). Application of sparse matrix methods in electric power system analysis. *In* "Sparse Matrix Proceedings" (R. A. Willoughby, ed.), Rep. No. RA 1(# 11707), pp. 113–121. IBM, Yorktown Heights, New York.

Chen, Y. T. (1972). Permutation of irreducible sparse matrices to upper triangular form. *IMA J.* **10**, 15–18.

Chen, Y. T., and Tewarson, R. P. (1972a). On the fill-in when sparse vectors are orthonormalized. *Computing (Arch. Elektron. Rechnen)* **9**, 53–56.

Chen, Y. T., and Tewarson, R. P. (1972b). On the optimal choice of pivots for the Gaussian elimination. *Computing* **9** (forthcoming).

Chen, W. K. (1967). On directed graph solution of linear algebraic equations. *SIAM Rev.* **9**, 692–707.

Churchill, M. E. (1971). A sparse matrix procedure for power system analysis programs. *In* "Large Sparse Sets of Linear Equations" (J. K. Reid, ed.), pp. 127–138. Academic Press, New York.

Clasen, R. J. (1966). Techniques for automatic tolerance control in linear programming. *Comm. ACM* **9**, 802.

Comstock, D. R. (1964). A note on multiplying Boolean matrices II. *Comm. ACM* **7**, 13.

Curtis, A. R., and Reid, J. K. (1971a). Fortran subroutines for the solution of sparse sets of linear equations. Rep. R 6844, Atomic Energy Res. Establishment, Harwell, England.

Curtis, A. R., and Reid, J. K. (1971b). On automatic scaling of matrices for Gaussian elimination. Rep. TP 444, Atomic Energy Res. Establishment, Harwell, England.

Curtis, A. R., and Reid, J. K. (1971c). The solution of large sparse systems of linear equations. Proceedings of IFIP Rep. TP 450, Atomic Energy Res. Establishment, Harwell, England.

Curtis, A. R., Powell, M. J. D., and Reid, J. K. (1972). On the estimation of sparse Jacobian matrices. Rep. TP 476, Atomic Energy Res. Establishment, Harwell, England.

Cuthill, E. H. (1971). Several strategies for reducing the bandwidth of matrices. Tech. note CMD-42-71, Naval Ship Res. and Develop. Center, Bethesda, Maryland.

Cuthill, E. H., and McKee, J. (1969). Reducing the bandwidth of sparse symmetric matrices. Tech. note AML-40-69. Appl. Math. Lab., Naval Ship Res. and Develop. Center, Washington, D.C.

Dantzig, G. B. (1963a). "Linear Programming and Extensions." Princeton Univ. Press, Princeton, New Jersey.

Dantzig, G. B. (1963b). Compact basis triangularization for the simplex method. *In* "Recent Advances in Mathematical Programming" (R. L. Graves and P. Wolfe, eds.), pp. 125–132, McGraw-Hill, New York.

Dantzig, G. B., Harvey, R. P., McKnight, R. D., and Smith, S. S. (1969). Sparse matrix techniques in two mathematical programming codes. *In* "Sparse Matrix Proceedings" (R. A. Willoughby, ed.), Rep. No. RA 1(# 11707), pp. 85–99. IBM, Yorktown Heights, New York.

Dantzig, G. B., and Orchard-Hays, W. (1954). The product form of inverse in the simplex method. *Math. Comput.* **8**, 64–67.

Dantzig, G. B., and Wolfe, P. (1961). The decomposition algorithm for linear programs. *Econometrica* **29**, 767–778.

Davis, P. J. (1962). Orthonormalizing codes in numerical analysis. *In* "Survey of Numerical Analysis" (J. Todd, ed.), pp. 347–379. McGraw-Hill, New York.

Dickson, J. C. (1965). Finding permutation operations to produce a large triangular submatrix. 28th Nat. Meeting of OR Society of America, Houston, Texas.

Douglas, A. (1971). Examples concerning efficient strategies for Gaussian elimination. *Computing* **8**, 382–394.

Duff, I. (1972). On a factored form of the inverse for sparse matrices. D.Phil. thesis, Oxford University.

Dulmage, A. L., and Mendelsohn, N. S. (1962). On the inversion of sparse matrices. *Math. Comput.* **16**, 494–496.

Dulmage, A. L., and Mendelsohn, N. S. (1963). Two algorithms for bipartite graphs. *SIAM J. Appl. Math.* **11**, 183–194.

Dulmage, A. L., and Mendelsohn, N. S. (1967). Graphs and matrices. *In* "Graph Theory and Theoretical Physics" (F. Harary, ed.), pp. 167–277. Academic Press, New York.

Edelmann, H. (1963). Ordered triangular factorization of matrices. *Proc. Power Systems Comput. Conf., London.*

Edelmann, H. (1968). Massnahmen zur Reduktion des Rechenaufwands bei der Berechnung grosser elektrischer Netze. *Elektron. Rechenanlagen* **10**, 118–123.

Erisman, A. M. (1972). Sparse matrix approach to the frequency domain analysis of linear passive electrical networks. *In* "Sparse Matrices and Their Application" (D. J. Rose and R. A. Willoughby, eds.), pp. 31–40. Plenum Press, New York.

Eufinger, J. (1970). Eine Untersuchung zur Auflösung magerer Gleichungssysteme. *J. Reine Angewandte Math.* **245**, 208–220.

Eufinger, J., Jaeger, A., and Wenke, V. K. (1968). An algorithm for the partitioning of a large system of sparse linear equations using graph theoretical methods. *In* "Methods of Operations Research" (R. Henn, H. P. Kunzi, H. Schubert, eds.), pp. 118–128. Verlag Anton Hain, Meisenheim, Germany.

Evans, D. J. (1972). New iterative procedures for the solution sparse systems of linear difference equations. *In* "Sparse Matrices and Their Applications" (D. J. Rose and R. A. Willoughby, eds.), pp. 89–100. Plenum Press, New York.

Faddeev, D. K., and Faddeeva, V. N. (1963). "Computational Methods of Linear Algebra." Freeman, San Francisco, California.

Forrest, J. J. H., and Tomlin, J. A. (1972). Updating triangular factors of the basis to maintain sparsity in the product form simplex method. *Math. Programming* **2**, 263–268.

Forsythe, G. E. (1967). Today's computational methods of linear algebra. *SIAM Rev.* **9**, 489–515.

Forsythe, G. E., and Moler, C. B. (1967). "Computer Solution of Linear Algebraic Systems." Prentice-Hall, Englewood Cliffs, New Jersey.

Fox, L. (1965). "Introduction to Numerical Linear Algebra." Oxford Univ. Press (Clarendon), London and New York.

Fulkerson, D. R., and Gross, O. A. (1965). Incidence matrices and interval graphs. *Pacific J. Math.* **15**, 835–855.

Fulkerson, D. R., and Wolfe, P. (1962). An algorithm for scaling matrices. *SIAM Rev.* **4**, 142–146.

Gass, S. (1958). "Linear Programming: Methods and Applications." McGraw-Hill, New York.

Gear, C. W. (1971). Simultaneous numerical solution of differential-algebraic equations, *IEEE Trans. Circuit Theory* **CT 18**, 89–95.

George, J. A. (1971). Computer implementation of the finite element method. Ph.D. thesis, Comput. Sci. Dept., Stanford Univ., Stanford, California.

George, J. A. (1972). Block elimination of finite element systems of equations. *In* "Sparse Matrices and Their Applications" (D. J. Rose and R. A. Willoughby, eds.), pp. 101–114. Plenum Press, New York.

Gibbs, N. E. (1969). The bandwidth of graphs. Ph.D. thesis, Purdue Univ.

Glaser, G. H. (1972). Automatic bandwidth reduction techniques. Rep. 72-260. DBA Systems Inc., Melbourne, Florida.

Glaser, G. H., and Saliba, M. S. (1972). Application of sparse matrices to analytical photogrammetry. *In* "Sparse Matrices and Their Applications" (D. J. Rose and R. A. Willoughby, eds.), pp. 135–146. Plenum Press, New York.

Gustavson, F. G. (1972). Some basic techniques for solving sparse systems of linear equations. *In* "Sparse Matrices and Their Applications" (D. J. Rose and R. A. Willoughby, eds.), pp. 41–52. Plenum Press, New York.

Gustavson, F. G., Liniger, W., and Willoughby, R. A. (1970). Symbolic generation of an optimal Crout algorithm for sparse systems of equations. *ACM J.* **17**, 87–109.

Guymon, G. L., and King, I. P. (1972). Application of the finite element method to regional transport phenomena. *In* "Sparse Matrices and Their Applications" (D. J. Rose and R. A. Willoughby, eds.), pp. 115–120. Plenum Press, New York.

Hachtel, G., Brayton, R., and Gustavson, F. (1971). The sparse tableau approach to network analysis and design. *IEEE Trans. Circuit Theory* **CT-18**, 101–113.

Hachtel, G., Gustavson, F., Brayton, R., and Grapes, T. (1969). A sparse matrix approach to network analysis. *Proc. Cornell Conf. Computerized Electron.*

Hadley, G. (1962). "Linear Programming." Addison-Wesley, Reading, Massachusetts.

Harary, F. (1959). A graph theoretic method for complete reduction of a matrix with a view toward finding its eigenvalues. *J. Math. Phys.* **38**, 104–111.

Harary, F. (1960). On the consistency of precedence matrices. *J. Assoc. Comput. Mach.* **7**, 255–259.

Harary, F. (1962). A graph theoretic approach to matrix inversion by partitioning. *Numer. Math.* **4**, 128–135.

Harary, F. (1967). Graphs and Matrices. *SIAM Rev.* **9**, 83–90.

Harary, F. (1969). "Graph Theory." Addison-Wesley, Reading, Massachusetts.

Harary, F. (1971a). Sparse matrices and graph theory. *In* "Large Sparse Sets of Linear Equations" (J. K. Reid, ed.), pp. 139–150. Academic Press, New York.

Harary, F. (1971b). Sparse digraphs: Classification and algorithms. Paper presented at *IFIP Conf., Ljubljana, Yugoslavia.*

Heap, B. R. (1966). Random matrices and graphs. *Numer. Math.* **8**, 114–122.

Hildebrand, F. B. (1956). "Introduction to Numerical Analysis." McGraw-Hill, New York.

Householder, A. S. (1958). Unitary triangularization of a nonsymmetric matrix. *J. ACM* **5**, 339–342.

Hsieh, H. Y., and Ghausi, M. S. (1971a). On sparse matrices and optimal pivoting algorithms. Tech. Rep. 400–213. Electrical Eng. Dept., New York Univ., New York.

Hsieh, H. Y., and Ghausi, M. S. (1971b). A probabilistic approach to optimal pivoting and prediction of fill-in for random sparse matrices. Tech. Rep. 400–214. Electrical Eng. Dept., New York Univ., New York.

Ingerman, P. Z. (1962). Path matrix, Algorithm 141. *Comm. ACM* **5**, 556.

Irons, B. M. (1970). A frontal solution program for finite element analysis. *Int. J. Numer. Methods Eng.* **2**, 5–32.

Jennings, A. (1966). A compact storage scheme for the solution of symmetric linear simultaneous equations. *Compt. J.* **9**, 281–285.

Jennings, A. (1968). A sparse matrix scheme for the computer analysis of structures. *Internat. J. Comput. Math.* **2**, 1–21.

Jennings, A., and Tuff, A. D. (1971). A direct method for the solution of large sparse symmetric simultaneous equations. *In* "Large Sparse Sets of Linear Equations" (J. K. Reid, ed.), pp. 97–104. Academic Press, New York.

Jensen, H. G. (1967). Efficient matrix techniques applied to transmission tower design. *Proc. IEEE* **55**, 1997–2000.

Jensen, H. G., and Parks, G. A. (1970). Efficient solutions for linear matrix equations. *J. Struct. Div. Proc. Amer. Soc. Civil Eng.* **96**, 49–64.

Jimenez, A. J. (1969). Computer handling of sparse matrices. Rep. No. TR 00.1873. IBM, Yorktown Heights, New York.

Kettler, P. C., and Weil, R. L. (1969). An algorithm to provide structure for decomposition. *In* "Sparse Matrix Proceedings" (R. A. Willoughby, ed.), Rep. No. RA1 (#11707), pp. 11–24. IBM, Yorktown Heights, New York.

King, I. P. (1970). An automatic reordering scheme for simultaneous equations derived from network analysis. *Internat. J. Numer. Methods Eng.* **2**, 523–533.

Klyuyev, V. V., and Kokovkin-Shcherbak, N. I. (1965). On the minimization of the number of arithmetic operations for the solution of linear algebraic systems of equations (translated by G. J. Tee). Rep. CS-24. Comput. Sci. Dept., Stanford University, Stanford, California.

Kron, G. (1963). "Diakoptics." McDonald, London.

Larson, L. J. (1962). A modified inversion procedure for product form of inverse in linear programming codes. *Comm. ACM* **5**, 382–383.

Lee, H. B. (1969). An implementation of Gaussian elimination for sparse systems of linear equations. *In* "Sparse Matrix Proceedings" (R. A. Willoughby, ed.), Rep. No. RA 1 (#11707), pp. 75–84. IBM, Yorktown Heights, New York.

Levy, R. (1971). Resequencing of the structural stiffness matrix to improve computational efficiency. *JPL Quart. Tech. Rev.* **1**, 61–70.

Liebl, P., and Sedlacek, J. (1966). Umformung von Quadratmatrizen auf quasitrianguläre Form mit Mitteln der Graphentheorie. *Appl. Mat.* **11**, 1–9.

Liniger, W., and Willoughby, R. A. (1969). Efficient numerical integration of stiff systems of differential equations. *SIAM J. Numer. Anal.* **7**, 47–66.

Livesley, R. K. (1960–1961). The analysis of large structural systems. *Comput. J.* **3**, 34–39.

Luksan, L. (1972). A collection of programs for operations involving sparse matrices. Res. Rep. Z-483. Inst. of Radio Eng. and Electronics, CSAV, Prague, Czechoslovakia.

Marimont, R. B. (1959). A new method for checking the consistency of precedence matrices. *J. Assoc. Comput. Mach.* **6**, 164–171.

Marimont, R. B. (1960). Application of graphs and Boolean matrices to computer programming. *SIAM Rev.* **2**, 259–268.

Marimont, R. B. (1969). System connectivity and matrix properties. *Bull. Math. Biophys.* **31**, 255–274.

Markowitz, H. M. (1957). The elimination form of the inverse and its application to linear programming. *Management Sci.* **3**, 255–269.

Maurser, W. D. (1968). "Programming, Introduction to Computer Languages and Techniques." Holden-Day, San Francisco, California.

Mayoh, B. H. (1965). A graph technique for inverting certain matrices. *Math. Comput.* **19**, 644–645.

McCormick, C. W. (1969). Application of partially banded matrix methods to structural analysis. *In* "Sparse Matrix Proceedings" (R. A. Willoughby, ed.), Report No. RA 1 (#11707), pp. 155–158. IBM, Yorktown Heights, New York.

McNamee, J. M. (1971). A sparse matrix package. Algorithm 408. *Comm. ACM* **14**, 265–273.

Mueller-Merbach, H. (1964). On round-off errors in Linear Programming. Res. Rep., Operations Res. Center, Univ. of California, Berkeley, California.

Nathan, A., and Even, R. K. (1967–1968). The inversion of sparse matrices by a strategy derived from their graphs. *Comput. J.* **10**, 190–194.

Norin, R. S., and Pottle, C. (1971). Effective ordering of sparse matrices arising from nonlinear electrical networks. *IEEE Trans. Circuit Theory* **CT-18**, 139–145.

Nuding, E., and Kahlert-Warmbold, I. (1970). A computer oriented representation of matrices. *Computing* **6**, 1–8.

Ogbuobiri, E. C. (1970). Dynamic storage and retrieval in sparsity programming. *IEEE Trans. Power Apparatus Systems* **PAS 89**, 150–155.

Ogbuobiri, E. C. (1971). Sparsity techniques in power-system grid-expansion planning. *In* "Large Sparse Sets of Linear Equations" (J. K. Reid, ed.), pp. 219–230. Academic Press, New York.

Ogbuobiri, E. C., Tinney, W. F., and Walker, J. W. (1970). Sparsity-directed decomposition for Gaussian elimination on matrices. *IEEE Trans. Power Apparatus Systems* **PAS 89**, 141–155.

Orchard-Hays, W. (1968). "Advanced Linear Programming Computing Techniques." McGraw-Hill, New York.

Orchard-Hays, W. (1969). MP systems technology for large sparse matrices. *In* "Sparse Matrix Proceedings" (R. A. Willoughby, ed.), Rep. No. RA1 (11707), pp. 59–64. IBM, Yorktown Heights, New York.

Palacol, E. L. (1969). The finite element method of structural analysis. *In* "Sparse Matrix Proceedings" (R. A. Willoughby, ed.), Report No. RA1 (11707), pp. 101–106. IBM, Yorktown Heights, New York.

Parter, S. (1960). On the eigenvalues and eigenvectors of a class of matrices. *J. SIAM* **8**, 376–388.

Parter, S. (1961). The use of linear graphs in Gauss elimination. *SIAM Rev.* **3**, 119–130.

Paton, K. (1971). An algorithm for the blocks and cut-nodes of a graph. *Comm. ACM* **14**, 468–475.

Pope, A. J., and Hanson, R. H. (1972). An algorithm for the pseudoinverse of sparse matrices. NOAA, Geodetic Research Lab., Rockville, Maryland. (Paper presented at spring A.G.U. meeting, Washington, D.C.)

Rabinowitz, P. (1968). Applications of linear programming to numerical analysis. *SIAM Rev.* **10**, 121–159.

Ralston, A. (1965). "A First Course in Numerical Analysis." McGraw-Hill, New York.

Reid, J. K. (ed.) (1971). "Large Sparse Sets of Linear Equations." *Proc. Oxford Conf. Inst. Math. Appl., April 1970.* Academic Press, New York.

Rice, J. H. (1966). Experiments on Gram-Schmidt orthogonalization. *Math. Comput.* **20**, 325–328.

Roberts, E. J. (1970). The fully indecomposable matrix and its associated bipartite graph —an investigation of combinatorial and structural properties. Rep. TM X-58037, NASA Manned Spacecraft Center, Houston, Texas.

Rogers, A. (1971). "Matrix Methods in Urban and Regional Analysis." Holden-Day, San Francisco, California.

Rose, D. J. (1970a). Symmetric elimination on sparse positive definite systems and the potential flow network problem. Ph.D. thesis. Harvard Univ.

Rose, D. J. (1970b). Triangulated graphs and the elimination process. *J. Math. Anal. Appl.* **32**, 597–609.

Rose, D. J. (1972). A graph-theoretic study of the numerical solution of sparse positive definite systems of linear equations. Math. Dept. Rep. University of Denver, Denver, Colorado.

Rose, D. J., and Bunch, J. R. (1972). The role of partitioning in the numerical solution of sparse systems. *In* "Sparse Matrices and Their Applications" (D. J. Rose and R. A. Willoughby, eds.), pp. 177–190. Plenum Press, New York.

Rose, D. J., and Willoughby, R. A., eds. (1972). "Sparse Matrices and Their Applications." *Proc. IBM Conf. Sept. 1970.* Plenum Press, New York.

Rosen, R. (1968). Matrix bandwidth minimization. *Proc. 23rd Nat. Conf. ACM Publ.* **P-68**, pp. 585–595. Brandon Systems Press, Princeton, New Jersey.

Ross, I. C., and Harary, F. (1959). A description of strengthening and weakening members of a group. *Sociometry* **22**, 139–147.

Roth, J. P. (1959). An application of algebraic topology: Kron's method of tearing. *Quart. Appl. Math.* **17**, 1–24.

Rubinstein, M. F. (1967). Combined analysis by substructures and recursion. *Proc. J. Struct. Div. ASCE* **93**, 231–235.

Rutishauser, H. (1963). On Jacobi rotation patterns. *Proc. Symposia Appl. Math.* **15**, pp. 219–240. Amer. Math. Soc., Providence, Rhode Island.

Sato, N., and Tinney, W. F. (1963). Techniques for exploiting the sparsity of the network admittance matrix. *IEEE Trans. Power Apparatus Systems*, **PAS-82**, 944–950.

Schubert, L. K. (1970). Modification of a quasi-Newton method for non-linear equations with sparse Jacobian. *Math. Comput.* **25**, 27–30.

Schwarz, H. R. (1968). Tridiagonalization of a symmetric band matrix. *Numer. Math.* **12**, 231–241.

Segethova, J. (1970). Elimination procedures for sparse symmetric linear algebraic systems of a special structure. Rep. No. 70–121. Comput. Sci. Center, Univ. of Maryland.

Smith, D. M. (1969). Data logistics for matrix inversion. *In* "Sparse Matrix Proceedings" (R. A. Willoughby, ed.), Report No. RA1 (11707), pp. 127–137. IBM, Yorktown Heights, New York.

Smith, D. M., and Orchard-Hays, W. (1963). Computational efficiency in product form LP codes. *In* "Recent Advances in Mathematical Programming" (R. L. Graves and P. Wolfe, eds.), pp. 211–218. McGraw-Hill, New York.

Spillers, W. R. (1968). Analysis of large structures: Kron's method and more recent work. *J. Struct. Div. ASCE 94* **ST-11**, 2521–2534.

Spillers, W. R., and Hickerson, N. (1968). Optimal elimination for sparse symmetric systems as a graph problem. *Quart. Appl. Math.* **26**, 425–432.

Stagg, G. W., and El-Abiad, A. H. (1968). "Computer Methods in Power System Analysis." McGraw-Hill, New York.

Steward, D. V. (1962). On an approach to techniques for the analysis of the structure of large systems of equations. *SIAM Rev.* **4**, 321–342.

Steward, D. V. (1965). Partitioning and tearing systems of equations. *SIAM J. Numer. Anal.* **2**, 345–365.

Steward, D. V. (1969). Tearing analysis of the structure of disorderly sparse matrices. *In* "Sparse Matrix Proceedings" (R. A. Willoughby, ed.), Rep. No. RA1 (11707), pp. 65–74. IBM, Yorktown Heights, New York.

Tewarson, R. P. (1966). On the product form of inverses of sparse matrices. *SIAM Rev.* **8**, 336–342.

Tewarson, R. P. (1967a). On the product form of inverses of sparse matrices and graph theory. *SIAM Rev.* **9**, 91–99.

Tewarson, R. P. (1967b). Solution of a system of simultaneous linear equations with a sparse coefficient matrix by elimination methods. *Nordisk. Tidskr. Informations Behandling (BIT)* **7**, 226–239.

Tewarson, R. P. (1967c). Row column permutation of sparse matrices. *Comput. J.* **10**, 300–305.

Tewarson, R. P. (1968a). On the orthonormalization of sparse vectors. *Computing (Arch. Elektron. Rechnen)* **3**, 268–279.

Tewarson, R. P. (1968b). Solution of linear equations with coefficient matrix in band form. *Nordisk. Tidskr. Informations Behandling (BIT)* **8**, 53–58.

Tewarson, R. P. (1969a). The Crout reduction for sparse matrices. *Comput. J.* **12**, 158–159.

Tewarson, R. P. (1969b). The Gaussian elimination and sparse systems. *In* "Sparse Matrix Proceedings" (R. A. Willoughby, ed.), Report No. RA1 (11707), pp. 35–42. IBM, Yorktown Heights, New York.

Tewarson, R. P. (1970a). On the transformation of symmetric sparse matrices to the triple diagonal form. *Internat. J. Comput. Math.* **2**, 247–258.

Tewarson, R. P. (1970b). Computations with sparse matrices. *SIAM Rev.* **12**, 527–544.

Tewarson, R. P. (1970c). On the reduction of a sparse matrix to Hessenberg form. *Internat. J. Comput. Math.* **2**, 283–295.

Tewarson, R. P. (1971). Sorting and ordering sparse linear systems. *In* "Large Sparse Sets of Linear Equations" (J. K. Reid, ed.), pp. 151–168. Academic Press, New York.

Tewarson, R. P. (1972). On the Gaussian elimination for inverting sparse matrices. *Computing (Arch. Elektron. Rechnen)* **9**, 1–7.

Tewarson, R. P., and Cheng, K. Y. (1972). A desirable form for sparse matrices when computing their inverses in factored forms *Computing* **9** (forthcoming).

Tinney, W. F. (1969). Comments on sparsity techniques for power system problems. *In* "Sparse Matrix Proceedings" (R. A. Willoughby, ed.), Report No. RA1 (11707), pp. 25–34. IBM, Yorktown Heights, New York.

Tinney, W. F., and Ogbuobiri, E. C. (1970). Sparsity techniques: theory and practice. March 1970 Rep. Bonnville Power Administration, Portland, Oregon.

Tinney, W. F., and Walker, J. W. (1967). Direct solutions of sparse network equations by optimally ordered triangular factorization. *Proc. IEEE* **55**, 1801–1809.

Tocher, J. L. (1966). Selective inversion of stiffness matrices. *Proc. Struct. Div. ASCE* **92**, 75–88.

Tomlin, J. A. (1970). Maintaining sparse inverse in the simplex method. Rep. 70-15. Operations Res. Dept., Stanford University, Stanford, California.

Tomlin, J. A. (1972a). Modifying triangular factors of the basis in the simplex method. *In* "Sparse Matrices and Their Applications" (D. J. Rose and R. A. Willoughby, eds.), pp. 77–85. Plenum Press, New York.

Tomlin, J. A. (1972b). Pivoting for size and sparsity in linear programming inversion routines. *IMA J* (forthcoming).

van der Sluis, A. (1969). Condition numbers and equilibration of matrices. *Numer. Math.* **14**, 14–23.

Varga, R. A. (1962). "Matrix Iterative Analysis." Prentice-Hall, Englewood Cliffs, New Jersey.

Warshall, S. (1962). A theorem on Boolean matrices. *J. ACM* **9**, 11–12.

Weaver, W., Jr. (1967). "Computer Programs for Structural Analysis." Van Nostrand–Reinhold, Princeton, New Jersey.

Weil, R. L., Jr. (1968). The decomposition of economic production systems. *Econometrica* **36**, 260–278.

Wenke, V. K. (1964). Praktische Anwendung linearer Wirtschafts-modelle. *Unternehmensforschung* **8**, 33–46.

Westlake, J. R. (1968). "A Handbook of Numerical Matrix Inversion and Solution of Linear Equations." Wiley, New York.

Wilkinson, J. H. (1965). "The Algebraic Eigenvalue Problem." Oxford Univ. Press, London and New York.

Willoughby, R. A. (ed.) (1969). "Sparse Matrix Proceedings," Rep. No. RA1 (11707). IBM, Yorktown Heights, New York.

Wolfe, P. (1965). Error in solution of linear programming problems. *In* "Error in Digital Computation" (L. B. Rall, ed.), vol. 2, pp. 271–284. Wiley, New York.

Wolfe, P. (1969). Trends in linear programming computations. *In* "Sparse Matrix Proceedings" (R. A. Willoughby, ed.), Rep. No. RA1 (11707), pp. 107–112. IBM, Yorktown Heights, New York.

Wolfe, P., and Cutler, L. (1963). "Experiments in linear programming" (R. L. Graves and P. Wolfe, eds.), pp. 211–218. McGraw-Hill, New York.

Ziewkiewicz, O. C. (1967). "The Finite Element Method in Structural and Continuum Mechanics." McGraw-Hill, New York.

Zollenkopf, K. (1971). Bi-Factorization: Basic computational algorithm and programming techniques. *In* "Large Sparse Sets of Linear Equations" (J. K. Reid, ed.), pp. 75–96. Academic Press, New York.

Author Index

Numbers in italics refer to the pages on which the complete references are listed.

153

Subject Index

157